Methods of Matrix Algebra

Methods of Matrix Algebra

Edited by:

Maria Catherine Borres

ARCLER PRESS

www.arclerpress.com

Methods of Matrix Algebra

Maria Catherine Borres

Arcler Press
2010 Winston Park Drive,
2nd Floor
Oakville, ON L6H 5R7
Canada
www.arclerpress.com
Tel: 001-289-291-7705
001-905-616-2116
Fax: 001-289-291-7601
Email: orders@arclereducation.com

ISBN: 978-1-77361-220-1 (Hardcover)

ABOUT THE EDITOR

Catherine is currently taking up Master of Arts in Education Major in Mathematics in Philippine Normal University – Manila. She is currently working as a Content Developer for Mathematics at the Affordable Private Education Center (APEC Schools).

TABLE OF CONTENTS

PREFACE

This book is designed to teach the thought and techniques of basic Matrix algebra as a precise mathematical subject. Besides computational adeptness, there is an emphasis on uptake definitions and theorems, as well as reading, understanding and creating proofs. A strictly logical organization, complete and exceedingly detailed proofs of every theorem, suggestion on techniques for reading and handwriting proofs, and a variety of challenging theoretical exercises evidence slowly provide the novice with the tools and certainty to be able to study other mathematical topics in a rigorous fashion.

In many respects this book is design for beginners. It combines an illustration hearted approach, well rounded explanations and wealth of practice opportunities. It makes a form friend for those students who are commencing a circuit in Matrix Algebra at school or for undergraduate series in Mathematics, Engineering or Physics.

Applications of matrices are found in different scientific fields. In every field of physics, including classical mechanics, optics, electromagnetism and electrodynamics, they are used to study physical phenomena, such as the motion of rigid bodies. In computer graphics, they are used to manipulate 3D ideal and project them onto a 2-dimensional screen. Matrix calculus generalizes classical analytical notions such as derivatives and exponentials to higher dimensions. Matrices are used in economics to describe systems of economic relationships.

Matrix calculations can be often performed with different techniques. Many problems can be solved by either direct algorithms or iterative approaches. For example, the eigenvectors of a square matrix can be obtained by discovery a sequence of vectors x_n converging to an eigenvector when n tends to infinity. Perhaps by providing a concise insight into where these concepts came from, how they were revealed, and how their uses in the past relate to how they are presently used and taught may offer students with the extra understanding they need to put these concepts to use, whether it be as measuring procedure, or as functions. With this better insight, one would be able to see the value in studying trigonometry as an element of mathematics, instead of an isolated unit from the subject.

This book aims to help the readers understand the wonderful world of Matrix Algebra, to enlighten them on how this subject having worth in real world problems.

Editor

CHAPTER 1

INTRODUCTION TO MATRICES

INTRODUCTION

The importance of study of matrices lies in the fact that many situation in both pure and applied mathematical involve rectangular arrays of numbers. In many branches of business, biological and social sciences, it is necessary to express and use a set of number in arranged in a rectangular array. For example, suppose a firm produces three types of Good and which it sells to two customers . The monthly sales of these goods (in hundreds) are given in the following table.

		Monthly sales of goods		
		G_1	G_2	G_3
Customers	C_1	9	4	5
	C_2	2	6	7

Ignoring table heading, we usually write this information more concisely as

$$\begin{bmatrix} 9 & 4 & 5 \\ 2 & 6 & 7 \end{bmatrix}$$

Which is an example of matrix? It has two rows and three columns. A formal of matrix as a rectangular array of numbers follows:

Definition. Let F *denote* the field of real or complex numbers. A rectangular array of m n elements $a_{ij} \in F, (1 \le i \le m, 1 \le j \le n)$ arranged in m rows and n columns and enclosed by square brackets such as

$$\begin{bmatrix} a_{11} & a_{12} & \cdots & a_{1j} & \cdots & a_{1n} \\ a_{21} & a_{22} & \cdots & a_{2j} & \cdots & a_{2n} \\ \vdots & \vdots & \cdots & \vdots & \cdots & \vdots \\ a_{i1} & a_{i2} & \cdots & a_{1j} & \cdots & a_{1n} \\ \vdots & \vdots & \cdots & \vdots & \cdots & \vdots \\ a_{m1} & a_{m2} & \cdots & a_{mj} & \cdots & a_{mn} \end{bmatrix}$$

is called an $m \times n$ Over the field F (or with entries from the fields F). Matrices are usually designated by capital letters. Instead of writing all the entries of a matrix as above, it is convenient to write the same in an abbreviated notation as $A = [A_{IJ}]_{mxn}$. Here a_{ij} denotes the element (or entry) in the ith rows and the jth column of the matrix A which has ***m rows*** and ***n column***. Thus the subscript i and j of the element a_{ij} of a matrix indicate respectively the rows and the column in which a_{ij} is located.

If a matrix A has m rows and n columns then A is said to be a rectangular matrix of order (or size) $(m \times n)$. In case $m = n, A$ is said to be a square matrix of order n.

If $A = [a_{ij}]_{m \times n,} [a_{i1}\, a_{i2}\, ...\, a_{in}]$ is a $1 \times n$ matrix and is called vector. Similarly, a single column

$$\begin{bmatrix} a_{i1} \\ a_{i1} \\ \vdots \\ a_{mj} \end{bmatrix}$$

is an $m x 1$ matrix and is called column vector.

Definition

An upper (lower) triangular matrix (or a triangular matrix) square matrix all of whose elements below (above) the main diagonal (running from upper left to lower right corner) are zero. Thus

$$A = \begin{bmatrix} a_{11} & a_{11} & \cdots & a_{1n} \\ 0 & a_{11} & \cdots & a_{1n} \\ \vdots & \vdots & \cdots & \vdots \\ 0 & 0 & \cdots & a_{nn} \end{bmatrix}$$

is an upper triangular matrix.

$a_{11}, a_{22} \ldots, a_{nn},$ are called elements of the main diagonal.

Definition. A square matrix, all of whose elements are zero except those in the main diagonal, is called a diagonal matrix

Thus, a square matrix $A = \left[a_{ij}\right]$ is a diagonal matrix if $a_{ij} = 0$ whenever $i \neq j$.

(Note that this does not say $a_{ij} \neq 0$ whenever $i = j$). An example of a diagonal matrix is

$$\begin{bmatrix} a & 0 & 0 \\ 0 & b & 0 \\ 0 & 0 & c \end{bmatrix}$$

Here $a,b,c \in F$ which may or may not be zero.

Definition. A square matrix of size n in which each element of the main diagonal is one and the same nonzero number $k \in F$ and all other elements are zero is called a scalar matrix of order n.

Thus

$$\begin{bmatrix} k & 0 & 0 & 0 \\ 0 & k & 0 & 0 \\ 0 & 0 & k & 0 \\ 0 & 0 & 0 & k \end{bmatrix}$$ is a scalar matrix of order 4.

Every Scalar matrix is a diagonal matrix

When each $k = 1$ (the identity element of F with respect to multiplication), the scalar matrix of order n is called the identity matrix and is denoted by $\boldsymbol{I}_n$. Thus

$$\boldsymbol{I}_3 = \begin{bmatrix} 1 & 0 & 0 \\ 0 & 1 & 0 \\ 0 & 0 & 1 \end{bmatrix}, \boldsymbol{I}_4 = \begin{bmatrix} 1 & 0 & 0 & 0 \\ 0 & 1 & 0 & 0 \\ 0 & 0 & 1 & 0 \\ 0 & 0 & 0 & 1 \end{bmatrix}$$

are identify matrices of order 3 and 4 respectively.

ALGEBRA OF MATRICES

Definition. (Equality of Matrices). Two $m x\ n$ matrices $A-[a_{ij}]$ and $B=[b_{ij}]$ are equal if and only if $a_{ij}=b_{ij}$ for each $j.$ When A and B are equal, we write $A=B.$

Definition. (Zero Matrix). A matrix, whose every element is zero, is called a **zero matrix**. If it has m row and n columns, we denote it by $0_{m\times n}$ or simply by 0 if there is no danger of ambiguity about the number of its rows and columns.

Definition. (Addition of Matrices). Two matrices are said to be conformable for addition when they have the same number of rows and the same number of columns. Thus if $A=[a_{ij}]$ and $B=[b_{ij}] are m\times n$ matrices over the same field F, then they can be added and their sum is the matrix

$$A+B=[a_{ij}+b_{ij}]$$

of order $m\times n.$ $That$ is, to find the sum $A+B$ of two matrices A and B of the same order we add their corresponding elements.

Definition. (Additive Inverse of a Matrix). Given an $m\times n$ matrix $A=[a_{ij}],$ we define $-A[-a_i].$Thus $-A is an\ \ m\times n\ \ matrix$ and by definition,

$$A+(-A)=[a_{ij}+(-a_{ij})]=0=[a_{ij}+(-a_{ij})]=(-A)+A.$$

Thus $-A$ is the additive inverse of A.

Also, $0+A=A+0, i.e., 0$ is the identity with respect to $+.$

Theorem. For any three $m xn$ matrices A ,B and $B and$ C, we have

$A+(B+C)=(A+B)+C$ (Associative Law for Matrix Addition)

Proof.

Let $A=[a_{ij}], B=[b_{ij}], C=[c_{ij}].$Then,

$A+(b+C)=[a_{ij}]+[b_{ij}+c_{ij}]$

$=[a_{ij}+(b_{ij}+c_{ij})],$

$=[(a_{ij}+b_{ij})+c_{ij}],$

$=[a_{ij}+b_{ij}]+[c_{ij}],$

Theorem. The set of all $m x n$ matrices (over a field F) is an abelian group with respect to addition.

Definition. (Subtraction of Matrices). If A and B are two matrices of the same order, we define $A-B\, as\, A-B=A+(-B).$ For example, let

$$A=\begin{bmatrix} 2 & 1 & -3 \\ -4 & 0 & 1 \end{bmatrix}, B=\begin{bmatrix} 1 & -1 & 2 \\ 0 & 1 & 0 \end{bmatrix}$$

Then $A-B=A+(-B)$

$$=\begin{bmatrix} 2 & 1 & -3 \\ -4 & 0 & 1 \end{bmatrix}+\begin{bmatrix} -1 & 1 & -2 \\ 0 & -1 & 0 \end{bmatrix}$$

$$=\begin{bmatrix} 2-1 & 1+1 & -3-2 \\ -4+0 & 0-1 & 1+0 \end{bmatrix}=\begin{bmatrix} 1 & 2 & -5 \\ -4 & -1 & 1 \end{bmatrix}$$

Definition. (Multiplication of a Matrix by a Scalar). Let A + $[a_{ij}]$ and $k\in F.$ Then we define the Scalar multiple $k\,A of\; A by\, k\, as\, K\, A=[ka_{ij}].i.e., kA$ is the matrix obtained by multiplying each element of $A by\, k.$

The additive inverse of a matrix A can also be defined as $-A=(-1)A,$ where 1 is the multiplicative identity of F.

Example 1. Let

$$A=\begin{bmatrix} 1 & -2 & 3 \\ 4 & 5 & -6 \end{bmatrix}, then\, 2A=\begin{bmatrix} 2 & -4 & 6 \\ 8 & 10 & -12 \end{bmatrix}$$

Theorem. Let A and B be $m x n$ matrices over a field F and let $a,b\in F.$ Then

- $(a+b)A \quad = aA+bA$
- $a(A+B) = aA+aB$
- $a(bA) \quad =(ab)A$
- $1A \quad = A$

Proof. (i) Let $A=[a_{ij}],\ B=[b_{ij}]$ Then

$(a+b)A=[(a+b)a_{ij}]$

$=[aa_{ij}+ba_{ij}]=[a_{ij}]+[ba_{ij}]$

$=[aa_{ij}+ba_{ij}]=[aa_{ij}]+[ba_{ij}]$

(ii). $a(A+B)=a[a_{ij}+b_{ij}]=[aa_{ij}+ab_{ij}]=[aa_{ij}]+[ab_{ij}]$

The other result follows similarly and are left as an exercise.

Definition. (Multiplication of Matrices)

Two matrices A and B are said to be conformable for the product AB if the number of columns in A is equal to the number of rows in B. Thus if A is an $m\times n$ matrix and B is an $n\times-p$ matrix, then the product $C=AB$ of the matrices A and B is an $m\times p\, matrix\, C$ whose entries are determined as follows:

Let $A=[a_{ij}]_{m\times m}$ and $B=[b_{ij}]_{n\times p}$ then $Ab=C, where$

$C=[c_{ij}]_{m\times p}$ and $c_{ij}=a_{i1}b_{ij}+a_{i2}b_{2j}+\ldots+a_{in}b_{nj}=\sum_{k=1}^{n}a_{jk}b_{kj}$

In order words, c_{ij} is the sum if the products of the elements of the ith row of A with the corresponding elements of the jth column of B i.e., c_{ij} is the dot products of the ith row vector of A with the jth column vector of B.

Example 2. Let

$$A=\begin{bmatrix} 1 & 0 & -1 \\ 2 & 9 & 7 \\ 5 & 3 & 0 \end{bmatrix},\ B=\begin{bmatrix} 6 & 1 \\ 0 & 4 \\ -2 & 3 \end{bmatrix}$$

$$AB=\begin{bmatrix} 1 & 0 & -1 \\ 2 & 9 & 7 \\ 5 & 3 & 0 \end{bmatrix}\begin{bmatrix} 6 & 1 \\ 0 & 4 \\ -2 & 3 \end{bmatrix}$$

$$=\begin{bmatrix} 1(6)+0(0)-1(-2) & 1(1)+0(4)-1(3) \\ 2(6)+9(0)+7(-2) & 2(1)+9(4)+7(3) \\ 5(6)+3(0)+0(-2) & 5(1)+3(4)+0(3) \end{bmatrix}=\begin{bmatrix} 8 & -2 \\ -2 & 59 \\ 30 & 17 \end{bmatrix}$$

Note that here A is a 3×3 , matrix, B is a 3×2 matrix and AB is 3×2 matrix. Moreover, the product BA is not defined because the number of columns of B is not equal to the number of row of A.

The commutative law for multiplication of matrices does not hold in general

That is, matrices A and $B, AB\neq BA$ even when both $AB\, and\, BA$ are defined

Example 3. Let

$$A=\begin{bmatrix} 1 & 2 \\ 3 & 4 \end{bmatrix} and\ B=-\begin{bmatrix} 1 & 1 \\ 0 & 2 \end{bmatrix}$$ Then

$$AB=\begin{bmatrix} 1(1)+2(0) & 1(1)+2(2) \\ 3(1)+4(0) & 3(1)+4(2) \end{bmatrix}=\begin{bmatrix} 1 & 5 \\ 3 & 11 \end{bmatrix}$$

$$\begin{bmatrix} 1(1)+1(3) & 1(2)+1(4) \\ 0(1)+2(3) & 0(2)+2(4) \end{bmatrix}=\begin{bmatrix} 4 & 6 \\ 6 & 8 \end{bmatrix}$$

Then $Ab\neq BA$

We may have $AB=0$ when neither $A=0\ nor\ B=0$.

Example 4. Let

$$A=\begin{bmatrix}1 & 2 & 0\\ 1 & 1 & 0\\ -1 & 4 & 0\end{bmatrix} and\, B=\begin{bmatrix}0 & 0 & 0\\ 0 & 0 & 0\\ 1 & 4 & 9\end{bmatrix}$$ Then

$$Ab=\begin{bmatrix}0 & 0 & 0\\ 0 & 0 & 0\\ 0 & 0 & 0\end{bmatrix}, but\, neither\;\; A=0\, nor\;\; B=0$$

For three matrices A, B, C, we may have $AB=AC(or\, BA=CA)$ with $B\neq C$. Thus cancellation law does not hold for multiplication of matrices.

Example 5. Let

$$A=\begin{bmatrix}1 & 2 & 0\\ 1 & 1 & 0\\ -1 & 4 & 0\end{bmatrix},\quad B=\begin{bmatrix}1 & 2 & 3\\ 1 & 1 & -1\\ 2 & 2 & 2\end{bmatrix} and\, C=\begin{bmatrix}1 & 2 & 3\\ 1 & 1 & -1\\ 1 & 1 & 1\end{bmatrix}$$ Then

$$AB=\begin{bmatrix}3 & 4 & 1\\ 2 & 3 & 2\\ 3 & 2 & -7\end{bmatrix} and\, AC=\begin{bmatrix}3 & 4 & 1\\ 2 & 3 & 2\\ 3 & 2 & -7\end{bmatrix}$$

Here $AB=AC\, but\;\; B\neq C$

Theorem. If the matrices A, B and C are conformable for the indicated sums and products then

$(i)\, A(BC)=(AB)C$ (Associate Law for Matrix Multiplication)

$(ii)\, A(B+C)=AC+BC$ (Left Distributive Law for Matrix Multiplication)

$(A+B)C=AC+BC$ (Right Distributive Law for Matrix Multiplication)

$(ii)K(AB)=(KA)\, B=A(kB)$, where $k\in F$.

Proof. (i) Let A (BC) is an $m\times q$ matrix and $(i,j)th$ element of $A(BC)$ is the sum of the product of elements of the ith rows of A and the corresponding elements if the jth column of BC. Now jth column of BC consist of elements.

$$\sum_{\mu=1}^{P} b_{i\mu} c_{\mu j}, \sum_{\mu}^{P} b_{2\mu} c_{mu}, \ldots, m\sum_{\mu=1}^{P} b_{n\mu} c_{\mu j}$$

Hence (i, j) *element of* $A(BC)$ *is*

$$a_{i1}\left(\sum_{\mu=1}^{P} b_{i\mu} c_{\mu j}\right) + a_{i2}\left(\sum_{\mu}^{P} b_{2\mu} c_{mu}\right) + \ldots + a_{in}\left(\sum_{\mu=1}^{P} b_{n\mu} c_{\mu j}\right)$$

$$= \sum_{\lambda=1}^{n} a_{i\lambda}\left(\sum_{\mu=1}^{P} b_{\lambda\mu} c_{\mu j}\right)$$

$$= \sum_{\lambda=1}^{n}\sum_{\mu=1}^{P} a_{i\lambda}(b_{\lambda\mu} c_{\mu j})$$

$$= \sum_{\lambda=1}^{n}\sum_{\mu=1}^{P} (a_{i\lambda} b_{\lambda\mu}) c_{\mu j}$$

(Associative Law in F)

$$= \sum_{\mu=1}^{P}\left(\sum_{\lambda=1}^{n} a_{i\lambda} b_{\lambda\mu}\right) C_{\mu j},$$

(Interchanging the summation symbols)

Hence $,(i, j)th$ element of $A(BC) = (i, j)th$ element of (AB) C which is true for each I and each j.

Hence A(BC) = (AB) C.

(ii) Here we assume that A = $[a_{ij}]$ *is an* $m \times n$ matrix and each of B = $[b_{ij}]$ and $C = [c_{ij}]$ is an $n \times [p$ matrix, so that $B + C$ is defined and is an $n \times p$ matric. Moreover, $A(B + C), AB and\ AC$ are also define and each of them is an $m \times p$ matrix . $AB + AC$ is also an $m \times p$ matrix.

Now $B + C = [(b_{\lambda j} + c_{\lambda j})]$

So $(i, j)th$ element of $A(B + C) = \sum_{\lambda=1}^{n} a_{i\lambda}(b_{\lambda j} + c_{\lambda j})$

$$=\sum_{\lambda=1}^{n}\left(a_{i\lambda}b_{\lambda j}+c_{\lambda j}\right)$$

$$=\sum_{\lambda=1}^{n}a_{i\lambda}b_{\lambda j}\sum_{\lambda=1}^{n}a_{i\lambda}c_{\lambda j}$$

$$=(i,j)th\,element\;of\;AB+(i,j)th\,element\,of\;AC$$

$$=(i,j)th\,element\,of\,(AB+AC)$$

We can prove similarly that $(A+B)C+AC+BC$

(iii) Suppose $A=\left[a_{ij}\right]$ is of order $m\,x\,n\,and\;B=\left[b_{ij}\right]$ is of order $n\times p$. Hence

$$K\left(\sum_{\lambda=1}^{n}a_{i\lambda}b_{\lambda j}\right)=\sum_{\lambda=1}^{n}(ka_{i\lambda})b_{\lambda j}=\sum_{\lambda=1}^{n}a_{ij}\left(kb_{\lambda j}\right),$$

Because the terms involved in the summations are all elements of the field in which commutative and associative laws hold. Hence

$$(i\;,j)th\,element\,of\;k(AB)=(i,j)th\,element\,of\,(kA)B$$

$$=(i,j)th\,element\,of\;A(KB)$$

So $k(AB)=(kA)B=A(KB)$

Definition. (Transpose of a Matrix). Let $A=\left[a_{ij}\right]$ be an $m\times n$ matrix interchanging rows columns of A. Thus $A^T=\left[b_{ij}\right]$, where $b_{ij}=a_{ij}$. In other words, $(i,j)th$ elemnt of $A^T=(j,i)th$ element of A.

Thus the first row of A^T is the first column of A, second row of A^T is the second column of A and so on.

Theorem. If the matrices A and B are conformable for the sum A + B and the product AB, then

(i) $(A\pm B)^T=A^T\pm B^T$

$(ii)\ \left(A^T\right)^T = A$

$(iii)(kA)^T = kA^T, \quad wheer\ k \in F$

$(iv)\ (AB)^T = B^T A^T$

Proof. **(i)** Let

$A = [a_{ij}]$ *and* $B = [b_{ij}]$ *be* $m \times n$ *matrices.Then* $A + B$ *is an* $m \times n$ matrix

and $(A+B)^T$ is an $n \times m$ matrix. Now A^T is an $n \times m$ matrix and B^T also is an $n \times m$ matrix. Therefore, $A^T + B^T$ is an $n\,xm$ *matrix*

Thus each of $(A+B)^T$ *and* $A^T + B^T$ *is an* $n\,x\,m$ *matrix*. Now

$(i,j)th$ element of $(A+B)^T = B^T$ is an $n\,x\,m$ matrix .Now

$(i,j)th$ *element of* $(A+B)^T = (j,i)th$ element of $A+B$

$= a_{ij} + b_{ji}$

Thus for all i and j.

$(i,j)th$ *element of* $(A+B)^T = (i,j)th$ element of $A^T + B^T$

Hence $(A+B)^T = A^T + B^T$

Similarly $(A-B)^T = (A+(-B))^T = A^T + (-B)^T = A^T - B^T$

We can similarly prove that

$(A_1 + A_2 + \ldots + A_n)^T = A_1^T + A_2^T + \ldots + A_n^2$

(ii) $(i,j)th$ element of $\left(A^T\right)^T = (j,i)th$ element of A^T

$= (i,j)th$ element of A.

Thus

$\left(A^T\right)^T = A$

(iii) $(i,j)th$ element of $(KA)^T = (j,i)th\ element\ of\ kA$

$= ka_{ji}$

$=(i,j)th\ elment\ of\ kA^T$

Hence

$(kA)^T = kA^T$

(iv) Let $A=\left[a_{ij}\right]$ be an $m \times n\ matrix\ and\ B=\left[b_{ij}\right] be\ an\ n \times p$ matrix.

Then Ab IS OF ORDER $m\ x\ p$ and so $(AB)^T$ is of order $p \times m.$

Now B^T is of order $p \times n\ and\ A^T$ is of order $n \times m.$ Hence $B^T A^T$ is of order $p \times m$

So

$(i,j)th$ element of $(AB)^T = (j,i)th\ element\ of\ AB$

$$= \sum_{\lambda=1}^{n} a_{j\lambda} b_{\lambda i}$$

Hence

$(AB)^T = B^T A^T$

This result can be generalized for the product of a finite number of matrices. $A_1, A_2 \dots\ , A_n.$ Thus

$(A_1 A_2 \dots A_N)^T = A_n^T \dots A_2^T A_1^T$

Definition. Let be a square matrix of order n. We define

$A^n = A\ .A\ .A \dots A(n\ factors)$

It can be proved by induction that, for all $m,n \in N$

$A^m . A^n = A^{m \times n}$

$\left(A^m\right)^n = A^{mn}$

We also define $A^0 = I_n$

Also, if A and B are conformable for multiplication then $\left(AB\right)^n \neq A^n\ B^n$ in general However, $\left(AB\right)^n = A^n\ B^n$ holds if $AB = BA$

Definition. A square matrix A for which $A^{k+1} = a, (k\, being\ \ a\ positive\, integer)$ is called a periodic matrix

If k is the least positive integer for which $A^{k+1} = A,$ then is said to be of period if $k = 1,$ so that $A^2 = A,\ then\ A\, is$ called an idempotent matrix.

Example 6. Consider

$$A = \begin{bmatrix} 2 & -2 & -4 \\ -1 & 3 & 4 \\ 1 & -2 & -3 \end{bmatrix}$$

$$A^2 = \begin{bmatrix} 2 & -2 & -4 \\ -1 & 3 & 4 \\ 1 & -2 & -3 \end{bmatrix}\begin{bmatrix} 2 & -2 & -4 \\ -1 & 3 & 4 \\ 1 & -2 & -3 \end{bmatrix}$$

$$= \begin{bmatrix} 2 & -2 & -4 \\ -1 & 3 & 4 \\ 1 & -2 & -3 \end{bmatrix} = A$$

Hence A is independent

Definition. A square matrix A for which $A^p = 0, (\, p\, being\, a\ positive\, integer),$ is called nilpotent. If is the least positive integer for which $A^p = 0,$ then A is said to be nilpotent of nil potency index p.

Example 7. Let

$$A = \begin{bmatrix} 1 & 1 & 3 \\ 5 & 2 & 6 \\ -2 & -1 & -3 \end{bmatrix} \text{Then}$$

$$A^2 = \begin{bmatrix} 1 & 1 & 3 \\ 5 & 2 & 6 \\ -2 & -1 & -3 \end{bmatrix}\begin{bmatrix} 1 & 1 & 3 \\ 5 & 2 & 6 \\ -2 & -1 & -3 \end{bmatrix}$$

$$= \begin{bmatrix} 0 & 0 & 0 \\ 3 & 3 & 9 \\ -1 & -1 & -3 \end{bmatrix}$$

$$A^3 = A^2 A = \begin{bmatrix} 0 & 0 & 0 \\ 3 & 3 & 9 \\ -1 & -1 & -3 \end{bmatrix}\begin{bmatrix} 1 & 1 & 3 \\ 5 & 2 & 6 \\ -2 & -1 & -3 \end{bmatrix}$$

$$= \begin{bmatrix} 0 & 0 & 0 \\ 0 & 0 & 0 \\ 0 & 0 & 0 \end{bmatrix} = 0$$

Hence A is nilpotent of nil potency index 3.

Definition. A square matrix A such *that* $A^2 = I$ is called an involuntary matrix

Example 8.

$$A = \begin{bmatrix} 1 & -1 & 0 \\ 0 & -1 & 0 \\ 0 & 0 & 1 \end{bmatrix} \text{Then}$$

$$A^2 = \begin{bmatrix} 1 & -1 & 0 \\ 0 & -1 & 0 \\ 0 & 0 & 1 \end{bmatrix}\begin{bmatrix} 1 & -1 & 0 \\ 0 & -1 & 0 \\ 0 & 0 & 1 \end{bmatrix} = \begin{bmatrix} 1 & 0 & 0 \\ 0 & 1 & 0 \\ 0 & 0 & 1 \end{bmatrix}$$

So that A is involuntary.

Definition. A square A for which $A^T = A$ is called a symmetric matrix A square matrix A for which $A^T = -A$ is called a skew symmetric matrix.

Note that the product of two symmetric matrices may not be symmetric. That is if $A^T - A,\ B^T = B$ then $(AB)^T = B^T A^T = BA \neq AB$ is general

Example 9. For

$$A = \begin{bmatrix} 1 & 2 & 3 \\ 2 & 4 & -3 \\ 3 & -3 & 6 \end{bmatrix},$$ we have $A^T = A.$ Hence A is symmetric.

Also for the matrix

$$A = \begin{bmatrix} 0 & -2 & 3 \\ 2 & 0 & 4 \\ -3 & -4 & 0 \end{bmatrix},$$ $A^T = -A$ Hence A is skew symmetric

Definition. Let $A = \left[a_{ij}\right]$ be a matrix with complex entries. Then the matrix $\left[a_{ij}\right]$, obtaine from A by replacing each a_{ij} by its complex conjugate, is called the conjugate matrix of A and is denoted by A (read A conjugate). $(A)^T$ is also denoted by A^H. Thus A is Hermitian if $A^H = A.$

A square matrix $A = \left[a_{ij}\right]$ over C for which $(A)^T = -A$ is called skew Hermitian.

As for the product of two symmetric matrices, the product of two Hermitian matrices A and B need not be Hermitian, That

$(AB)^H = B^H A^H = BA \neq AB\, in\, general$.

However, if A and B are conformable for multiplication then

$(B^T AB)^T = B^T A^T B\, and \left(B^H AB\right)^H = B^H A^H B.$

So, for symmetric (Hermitian) $A, B^T AB, \left(B^H AB\right)$ is symmetric (Hermitian).

Example 10. Let

$$A = \begin{bmatrix} 1 & 1-i & 2 \\ 1+i & 3 & i \\ 2 & -i & 0 \end{bmatrix}$$ Then $$A = \begin{bmatrix} 1 & 1+i & 2 \\ 1-i & 3 & -i \\ 2 & i & 0 \end{bmatrix}$$

and $$\left(A\right)^T = \begin{bmatrix} 1 & 1-i & 2 \\ 1+i & 3 & i \\ 2 & -i & 0 \end{bmatrix} = A. Thus\ A$$ is Hermitian.

Example 11. Let

$$A = \begin{bmatrix} 1 & 1-i & 2 \\ -1-i & 3i & i \\ -2 & i & 0 \end{bmatrix}$$ it is easy to verify that $\left(A\right)^T = -A.$

Hence A is skew Hermitian

Example 12. Show that for any square matrix $A, +A + A^T$ is symmetric and $A - A^T$ is skew symmetric.

Solution. Here we use the results $\left(A \pm B\right)^T = A^T \pm B^T\ and \left(A^T\right)^T = A.$ Thus

$$\left(A + A^T\right)^T = A^T + \left(A^T\right)^T = A^T + A = A + A^T$$

Hence $A + A^T$ is symmetric

Again $\left(A - A^T\right)^T = A^T - \left(A^T\right)^T = A^T - A = -\left(A - A^T\right).$

Hence $A - A^T$ is skew symmetric.

Analogous to Theorem, we have the following relations between two matrices A and B and their Hermitian forms.

Theorem. Let A and B be the square matrices of the same order. Then

$(i)\ \left(A^H\right)^H = A,\qquad (ii)\ (A \pm B)^H = A^H \pm B^H$

$(iii)\ (cA)^H = (c)A^H,\quad (iv)\qquad (AB)^H = B^H A^H.$

PARTITIONING OF MATRICES

Definition. Some time we partition a matrix into blocks of elements and consider such blocks as elements of the matrix. These blocks are called sub matrices of the original matrix and are labeled in the same manner as elements of a matrix. A matrix can be partitioned into sub matrices in many different ways and the manner in which it is to be partitioned is often indicated by drawing horizontal and vertical lines between selected rows and columns.

$$A = \begin{bmatrix} a_{11} & a_{12} \cdots & a_{1n} \\ a_{21} \quad a_{22} & \cdots & a_{2n} \\ \vdots & \vdots \;\cdots & \vdots \\ a_{m1} & a_{m2} \cdots & a_{mn} \end{bmatrix}$$

Let

Be an $m \times n$ matrix. We can write a partitioning of A as

$$A = \begin{bmatrix} B_{11} & B_{12} \\ B_{21} & B_{22} \end{bmatrix}, where$$

$$B_{11} = \begin{bmatrix} a_{11} & a_{12} \cdots & a_{1k} \\ a_{21} \quad a_{22} & \cdots & a_{2k} \\ \vdots & \vdots \;\cdots & \vdots \\ a_{p1} & a_{p2} \cdots & a_{pk} \end{bmatrix}$$

$$B_{12} = \begin{bmatrix} a_{1\,k+1} & a_{1\,k+2} \cdots & a_{1n} \\ a_{2\,k+1} \quad a_{2\,k+2} & \cdots & a_{2n} \\ \vdots & \vdots \;\cdots & \vdots \\ a_{p\,k+1} & a_{p\,K+2} \cdots & a_{pn} \end{bmatrix}$$

$$B_{21} = \begin{bmatrix} a_{p+11} & a_{p+12} \cdots & a_{p+1k} \\ a_{P+21} & a_{P+22} & \cdots & a_{p+2k} \\ \vdots & \vdots & \cdots & \vdots \\ a_{m1} & a_{m2} & \cdots & a_{mk} \end{bmatrix}$$

$$B_{22} = \begin{bmatrix} a_{p+1k+1} & a_{p+1K+2} \cdots & a_{p+1n} \\ a_{P+2\ K+1} & a_{P+2\ K+2} & \cdots & a_{p+2n} \\ \vdots & \vdots \ \cdots & \vdots \\ a_{m\,k+1} & a_{pm\,K+2} & \cdots & a_{mn} \end{bmatrix}$$

Here

B_{11} is a $P \times k$ matrix.

B_{12} is a $p \times (n-k)$ matrix

B_{21} is an $(m-p) \times k$ matrix

B_{22} is an $(m-p) \times (n-k)$ matrix.

Let A, B be two $m \times n$ matrices A and B are said to be identically partitioned if the corresponding sub matrices C_{ij} and D_{ij} of A B are of the same order (i.e have the same size). In such a case $C_{ij} \pm D_{ij}$ are both defined?

Example 13. Consider

$$A = \begin{bmatrix} 1 & -2 & 5 \\ 3 & 0 & 2 \\ 4 & 2 & -6 \end{bmatrix}$$

A can be partitioned as shown below

$$A = \begin{bmatrix} 1 & -2 & 5 \\ 3 & 0 & 2 \\ 4 & 2 & 6 \end{bmatrix}$$

$= \begin{bmatrix} A_{11} & A_{12} \\ A_{21} & A_{22} \end{bmatrix}$, where

$$A_{11} = \begin{bmatrix} 1 & -2 \\ 3 & 0 \end{bmatrix} is\, a\, 2 \times 2\, submatrix$$

$$A_{12} = \begin{bmatrix} 5 \\ 2 \end{bmatrix} is\, a\, 2 \times 1\, submatrix$$

$$A_{21} = \begin{bmatrix} 4 & 2 \end{bmatrix} is\, a_1 \times 2\, submatrix$$

$$A_{22} = \begin{bmatrix} -6 \end{bmatrix} is\ a_1 \times 1\, submatrix$$

Example 14. Partitioning of the matrices A and B are given as under.

$$A = \begin{bmatrix} 1 & 2 & 5 \\ -4 & 3 & 0 \\ 2 & 1 & 5 \end{bmatrix} = \begin{bmatrix} A_{11} & A_{12} \\ A_{21} & A_{22} \end{bmatrix}$$

And $B = \begin{bmatrix} a & b & c \\ d & e & f \\ g & h & i \end{bmatrix} = \begin{bmatrix} B_{11} & B_{12} \\ B_{21} & B_{22} \end{bmatrix}$

Here A and B are identically portioned.

$A_{11}\, and\, B_{11} are 1 \times 2$ submatrices

$A_{12}\, and\, B_{12} are 1 \times 1$ submatrices

$A_{21}\, and\, B_{21} are 2 \times 2$ submatrices

$A_{22}\, and\, B_{22} are 2 \times 1$ submatrices

Moreover, $A_{11} + B_{11}, A_{12} + B_{12}, A_{21} + B_{21}\, and\, A_{22} + B_{22}$ are all defined.

Example 15. Let

$$A = \begin{bmatrix} 1 & 0 & 0 & 0 & 0 & 1 \\ 0 & 1 & 0 & 0 & 0 & 1 \\ 0 & 0 & 1 & 0 & 0 & 1 \\ 0 & 0 & 0 & 1 & 0 & 0 \\ 0 & 0 & 0 & 0 & 1 & 0 \\ 1 & 1 & 1 & 0 & 0 & 1 \end{bmatrix}$$

We partition A as shown below

$$A = \begin{bmatrix} 1 & 0 & 0 & 0 & 0 & 1 \\ 0 & 1 & 0 & 0 & 0 & 1 \\ 0 & 0 & 1 & 0 & 0 & 1 \\ 0 & 0 & 0 & 1 & 0 & 0 \\ 0 & 0 & 0 & 0 & 1 & 0 \\ 1 & 1 & 1 & 0 & 0 & 1 \end{bmatrix}$$

$$= \begin{bmatrix} I_3 & 0_{3\times2} & A_1 \\ 0_{2\times3} & I_2 & 0_{2\times1} \\ A_1^T & 0_{1\times2} & [1] \end{bmatrix}, where\ A_1 = \begin{bmatrix} 1 \\ 1 \\ 1 \end{bmatrix}$$

Now

$$A^2 = \begin{bmatrix} I_3 & 0_{3\times2} & A_1 \\ 0_{2\times3} & I_2 & 0_{2\times1} \\ A_1^T & 0_{1\times2} & [1] \end{bmatrix} \begin{bmatrix} I_3 & 0_{3\times2} & A_1 \\ 0_{2\times3} & I_2 & 0_{2\times1} \\ A_1^T & 0_{1\times2} & [1] \end{bmatrix}$$

$$= \begin{bmatrix} I_3 + A_1A_1^T & 0_{3\times2} & A_1 + A_1 \\ 0_{2\times3} & I_2 & 0_{2\times1} \\ A_1^T + A_1^T & 0_{1\times2} & A_1^T A_1 + [1] \end{bmatrix}$$

$$I_3 + A_1A_1^T = I_3 + \begin{bmatrix} 1 \\ 1 \\ 1 \end{bmatrix} \begin{bmatrix} 1 & 1 & 1 \end{bmatrix}$$

$$= l_3 + \begin{bmatrix} 1 & 1 & 1 \\ 1 & 1 & 1 \\ 1 & 1 & 1 \end{bmatrix} = \begin{bmatrix} 2 & 1 & 1 \\ 1 & 2 & 1 \\ 1 & 1 & 2 \end{bmatrix}$$

$$A_1 + A_1 \quad = \begin{bmatrix} 2 \\ 2 \\ 2 \end{bmatrix}$$

$$A_1^T + A_1^T = [2 \;\; 2 \;\; 2],\; A_1^T A_1 + [1] = [3] + [1] = [4]^1 = 4$$

$$A^2 = \begin{bmatrix} 2 & 1 & 1 & 0 & 0 & 2 \\ 1 & 2 & 1 & 0 & 0 & 2 \\ 1 & 1 & 2 & 0 & 0 & 2 \\ 0 & 0 & 0 & 1 & 0 & 0 \\ 0 & 0 & 0 & 0 & 1 & 0 \\ 2 & 2 & 2 & 0 & 0 & 4 \end{bmatrix}$$

Hence

Note. In a number of cases dealing with matrices having bigger orders, the operations on matrices are done by partitioning them into matrices of small orders, This is done specially in the situation when the memory capacity of computer in limited. The size of the sub matrices must conform to the matrix operations to be done on the sub matrices.

Since each matrix always indicates some information partitioning of a matrix leads to the division of the information into separate blocks or parts. The same process can be reversed by taking into account the given matrices and adjoin them together and get a matrix of a bigger size which gives the information in consolidated form.

The partition of matrices is useful in a number of ways. Since matrices present information in a compact form, the same information could be divided into a number of interconnected sub-system to make the computation and calculation easier.

ELEMENTARY ROW OPERATIONS

Consider the identity matrix

$$l_3 = \begin{bmatrix} 1 & 0 & 0 \\ 0 & 1 & 0 \\ 0 & 0 & 1 \end{bmatrix}$$

The matrices

$$E_1 = \begin{bmatrix} 1 & 0 & 0 \\ 0 & 0 & 1 \\ 0 & 0 & 1 \end{bmatrix}, E_2 = \begin{bmatrix} 1 & 0 & 0 \\ 0 & 3 & 0 \\ 0 & 0 & 1 \end{bmatrix}, E_3 = \begin{bmatrix} 1 & 0 & 0 \\ 0 & 1 & 0 \\ 0 & -5 & 1 \end{bmatrix}$$

Have been obtained form l_3 by the following operations.

E_1 : Interchange of two rows of l_3. (second and third rows interchanged

E_2; Multiply a row of l_3 by a nonzero number. (Here row is multiplied by 3)

E_3 : Add a multiple of a row of l_3 to another row (-5 times row two is added to row three).

These operations on l_3 are called elementary row operations and corresponding matrices E_1, E_2, E_3 are called elementary matrices of order 3.

Now let

$$A = \begin{bmatrix} a_{11} & a_{12} & a_{13} \\ a_{21} & a_{22} & a_{23} \\ a_{31} & a_{32} & a_{33} \end{bmatrix}$$

We examine the effect of pre multiplying A by $E_1, E_2 \text{ and } E_3$ respectively

$$E_1 A = \begin{bmatrix} 1 & 0 & 0 \\ 0 & 0 & 1 \\ 0 & 1 & 0 \end{bmatrix} \begin{bmatrix} a_{11} & a_{12} & a_{13} \\ a_{21} & a_{22} & a_{23} \\ a_{31} & a_{32} & a_{33} \end{bmatrix} = \begin{bmatrix} a_{11} & a_{12} & a_{13} \\ a_{31} & a_{32} & a_{33} \\ a_{21} & a_{22} & a_{23} \end{bmatrix}$$

= A matrix obtained from A by interchanging of rows two and three

$$E_2A = \begin{bmatrix} 1 & 0 & 0 \\ 0 & 3 & 0 \\ 0 & 0 & 1 \end{bmatrix} \begin{bmatrix} a_{11} & a_{12} & a_{13} \\ a_{21} & a_{22} & a_{23} \\ a_{31} & a_{32} & a_{33} \end{bmatrix} = \begin{bmatrix} a_{11} & a_{12} & a_{13} \\ 3a_{21} & 3a_{22} & 3a_{23} \\ a_{31} & a_{32} & a_{33} \end{bmatrix}$$

$$E_3A = \begin{bmatrix} 1 & 0 & 0 \\ 0 & 1 & 0 \\ 0 & -5 & 1 \end{bmatrix} \begin{bmatrix} a_{11} & a_{12} & a_{13} \\ a_{21} & a_{22} & a_{23} \\ a_{31} & a_{32} & a_{33} \end{bmatrix} = \begin{bmatrix} a_{11} & a_{12} & a_{13} \\ a_{21} & a_{22} & a_{23} \\ a_{31}-5a_{21} & a_{32}-5a_{22} & a_{33}-5a_{23} \end{bmatrix}$$

$= A$ Matrix obtained from A by subtracting 5 times the second row from the third row.

If we post multiply the matrix A by

$$E_1' = \begin{bmatrix} 1 & 0 & 0 \\ 0 & 0 & 1 \\ 0 & 1 & 0 \end{bmatrix}, E_2' = \begin{bmatrix} 1 & 0 & 0 \\ 0 & 3 & 1 \\ 0 & 0 & 1 \end{bmatrix}, E_3' = \begin{bmatrix} 1 & 0 & 0 \\ 0 & 1 & -5 \\ 0 & 0 & 1 \end{bmatrix}$$

AE_1' , AE_2' *and* AE_3' are the matrices obtained from A by doing similar operation in columns of A instead of rows of A. This fact can be generalized to any $m \times n$ matrix.

That is the same operations can be performed on any arbitrary matrix

$$A = \begin{bmatrix} a_{11} & a_{12} & \cdots & a_{1n} \\ a_{21} & a_{22} & \cdots & a_{2n} \\ \vdots & \vdots & \vdots & \vdots \\ a_{m1} & a_{m2} & \cdots & a_{mn} \end{bmatrix}$$

By pre multiplying A by the $m \times m$ elementary matrices E_1, E_2 and E_3 obtained from I_m by the row operations mentioned above respectively.

Definition. The following operations on a matrix A are called elementary row operations:

(i) interchanging of any two rows $(R_{ij}$ denotes the interchanging of the ith row with the jth row)

(ii) Multiplication of a row of A by any n nonzero real or complex number $(kR_i$ denote that i th row is multiplied by $k \neq 0)$.

(iii) Addition of a scalar multiple of one row to another row. $(R_j + kR_i)$ denotes that k times row i is added to the j th row

Note that each of these row operation on an $m\times n$ matrix can be effected pre-multiplying A by an $m\times n$ elementary matrix.

Let A be an $m\ xn$ matrix. An $m \times n$ matric B is called row equivalent to $A if\ B$ obtained from A by performing a finite sequence of elementary row operations on A. We write $B\tilde{R}A$ to denote B is row equivalent to A.

Definition. An $m \times n\ matrix\ A$ is said to be in (row) form (or an echelon matrix)

If it has following properties.

(i) All nonzero rows are above any zero rows (consisting of all zero)

(ii) The first nonzero entry in each nonzero row is to the right of the first nonzero entry of each preceding row. That is, the number of zeros occurring before the first nonzero entry in each nonzero row is greater than the number of zeros that appear before the first nonzero element in any preceding row.

In an echelon matrix, the first nonzero entry of a row is called a pivot (or a row leader).There is at the most one pivot in each roe and in each column of an echelon matrix. A column containing a pivot is called a pivot column.

An echelon matrix in which each pivot is I and every other entry of the pivot column is zero. Is said to be in row reduced echelon form.

The matrices $\begin{bmatrix} 0 & 5 & 2 & 1 \\ 0 & 0 & 9 & -1 \\ 0 & 0 & 0 & 0 \end{bmatrix}$ and $\begin{bmatrix} 1 & 3 & 0 & 0 \\ 0 & 0 & 1 & 0 \\ 0 & 0 & 0 & 1 \\ 0 & 0 & 0 & 0 \end{bmatrix}$ are in echelon form. The second matrix is in row reduced echelon form.

Theorem. Every matrix is row equivalent to a matrix in echelon form (reduced echelon form)

Proof. Let A b a given $m\times n$ matrix. Either every element in the first column of A is zero or there exist a nonzero element a (say) In the th row, of this column of A is second case, interchanging the first and k th rows of A we obtain a matrix B whose first element $b_{11} = a \neq 0.$ Multiplying the element of the first row of C is other rows. The remaining elements of the first column can be made zeros. In the first case, we consider the second column and so on till we find a column with a nonzero entry. We then repeat the process as for the second case. Thus is row equivalent to a matrix in either of the forms.

$$\begin{bmatrix} 0 & C \end{bmatrix} \quad or \quad \begin{bmatrix} 1 & D \\ 0 & E \end{bmatrix}$$

In the second case 0 represent the $(m-1)\times 1$ *zero matrix*, $D\, a_1 \times (n-1)$ matrix and E an $(m-1)\times(n-1)\, matrix$. In the first case repeat the above process with the matrix C, whereas in the second case repeat the process with the matrix E. A continuation of this process leads to the desired form.

When a matrix is in echelon form, then by adding suitable multiplies of second, third,.... rows to the first, second, ... rows in succession, we get the reduced echelon form.

Example 16. Reduce the matrix

$A = \begin{bmatrix} 6 & 3 & -4 \\ -4 & 1 & -6 \\ 1 & 2 & -5 \end{bmatrix}$ in to the echelon form.

Solution. Here

$$A \tilde{R} \begin{bmatrix} 1 & 2 & -5 \\ -4 & 1 & -6 \\ 1 & 3 & -4 \end{bmatrix} \; by\, R_{13}$$

$$\tilde{R} \begin{bmatrix} 1 & 2 & -5 \\ 0 & 9 & -26 \\ 0 & -9 & 26 \end{bmatrix} \; by\, R_2 + 4R_1 \; and \; R_3 - 6R_1$$

$$\tilde{R} \begin{bmatrix} 1 & 2 & -5 \\ 0 & 9 & -26 \\ 0 & 0 & 0 \end{bmatrix} by \; R_3 + R_2$$

$$\tilde{R}\begin{bmatrix} 1 & 2 & -5 \\ 0 & 1 & -\frac{26}{9} \\ 0 & 0 & 0 \end{bmatrix} by \frac{1}{9}R_2$$

Which is the desired echelon form.

Corollary. A square matrix A is nonsingular if an only if it can be written as a product of elementary matrices

$$A = E_r E_{r-1} \ldots E_2 E_1$$

Proof. Here $A = \left(A^{-1}\right)^{-1}$ and by Theorem, we have

$$\left(A^{-1}\right) = E_r E_{r-1} \ldots E_2 E_1 I$$

$$= E_r E_{r-1} \ldots E_2 E_1$$

Hence $A = \left(E_r E_{r-1} \ldots E_2 E_1\right)^{-1}$

$$= E_1^{-1} E_2^{-1} \ldots E_{r-1}^{-1} E_r^{-1}$$

Since the inverse of an elementary matrix is an elementary matrix, A is said product of elementary matrices.

Conversely, suppose that A is the product of elementary matrices. Since each elementary matrix is nonsingular. Hence A is nonsingular.

Corollary. $B \widetilde{R} A$ if an only if $B = PA,$ where P is nonsingular.

ELEMENTARY COLUMN OPERATION

So far, we have been considering elementary row operation. We can consider elementary column operations in an analogous manner

Definition. The following operations on a matrix A are called elementary column operations.

- Interchanging of any two columns of
- Multiplication of a columns of A by a nonzero scalar.
- Addition of a scalar multiple of one column of A to another column

On applying an elementary column operation to the identify matrix, we obtain a corresponding elementary matrix, which is nonsingular.

Definition. An $m \times n$ matrix B is called column equivalent to an $m \times n$ matrix A if B is obtained by performing a sequence of a finite number of elementary column operations on A. We write $B \tilde{C} A$ to denote is column equivalent to A.

An elementary column operation on a matrix becomes an elementary rp operation on the transpose A^T *of* A. *Thus if* $B \tilde{C} A$, *then* $B^T \ \tilde{C} A^T$. Hence $B^T = EA^T$ where E is an elementary matrix. Thus $B = AE^T$ and we have proved.

Theorem. An elementary column operation on $m \times n$ matrix A amounts to post-multiplying A by an $n \times n$ elementary matrix corresponding to the elementary column operation.

Theorem. An $m \times n$ matrix B is column equivalent to an $m x n$ matrix A if and only if $B = AQ$, where Q is an $n x n$ elementary matrix.

Proof. Let $B \tilde{C} A$. Then $B^T \ \tilde{R} A^T$. *Hence* $B^T = PA^T$, *wher P is an* $n \times n$ nonsingular matrix. Now

$B = \tilde{C} A. \Rightarrow B = AQ$ for a nonsingular matrix

$B^T = (AQ)^T = Q^T A^T$ so that $B^T \tilde{R} A^T$ that is $B \tilde{C} A$.

Definition. If an $m \ x \ n$ matrix B can be obtained from an $m x n$ matrix A by a finite number of elementary row and column operations, then B is said to be equivalent to A, We write $B \sim A$ *to* denote B is equivalent to A.

The row and column equivalence are special cases of the general concepts of equivalence of matrices combining our previous results, we have

Theorem. An $m \times n$ matrix B is equivalent to an $m \times n$ matrix if A and only if $B = PAQ$ where P and Q are nonsingular and of order $m \times m$ and $n \times n$ respectively.

Theorem. Every nonzero mxn matrix is equivalent to n $m \times n$ matrix D where

$$D = \begin{bmatrix} I & 0 \\ 0 & 0 \end{bmatrix}$$

is called the canonical (or normal form) of the matrix $A)$

Proof. Since $A \neq 0,$ there exists a nonzero element a of A. By performing elementary roe and column operation on $A,$ we obtain an equivalent matrix with element in the (1,1) position.Applying the operations $A^{-1}R_1$, we get 1 in the (1,1) position. By adding suitable multiples of the first row to the remaining rows, every other element pf the first column can be made zero. Having done that, we can reduce every element of the first row except that in the (1, 1) position, to a zero. We thus obtain an equivalent matrix of the form

$$B = \begin{bmatrix} 1 & 0 \\ 0 & C \end{bmatrix}$$

Here *0* in the first row is a $1 \times (n-1)$ zero matrix, in the first column is an $(m-1) \times 1$ zero matrix and C is an $(m-1) \times (n-1)$ matrix. Repeating the process with $C,$ we get the required form.

HOMOGENIOUS SYSTEMS

Homogeneous systems play a central in linear algebra. This will be seen in where the foundations of the subject are all integrated to solve one of the major problems occurring in a wide variety of applications. In this section we deal with several problems involving homogeneous systems that will be fundamental in. Here we are able to focus our attention on these problems without being distracted by the additional material in.

Consider the homogeneous system

$Ax = 0.$

Where $A is an m \times n$ matrix. As we have already observed in of the set of all solutions to this homogeneous system is a subspace of R^n. An extremely important problems. Which will occur repeatedly in is that of finding a basis for this solution space. To find such a basis, we use the method of Gauss-Jordan reduction presented in Thus we transform the augmented matrix $[A \vdots 0]$. Of the system to a matrix $[B \vdots 0]$ in reduced row echelon form, where has nonzero rows , $1 \le r \le m$. Without loss of generality we may assume that the leading $1's$ in the nonzero rows of B occur in the first columns .if $r = n$. Then

$$[B \vdots 0] = \left[\begin{array}{cccccc|c} 1 & 0 & \dots & 0 & 0 & & 0 \\ 0 & 1 & \dots & 0 & 0 & & 0 \\ \vdots & \vdots & \ddots & \vdots & \vdots & & \vdots \\ 0 & 0 & \dots & 0 & 1 & & 0 \\ 0 & 0 & \dots & & 1 & & 0 \\ \vdots & \vdots & & & \vdots & & \vdots \\ 0 & 0 & \dots & & 0 & & 0 \end{array}\right] r = n\}m$$

And the only solution to $Ax = 0$ is the trivial one. The solution space has no basis and its dimension is zero.

If $r<n$, then

$$[B \vdots 0] = \left[\begin{array}{ccccccccc|c} 1 & 0 & 0 & \dots & 0 & b_{1r+1} & \dots & b_{1n} & & 0 \\ 0 & 1 & 0 & \dots & 0 & b_{2r+1} & \dots & b_{2n} & & 0 \\ 0 & 0 & 1 & \dots & 0 & \vdots & & \vdots & & \vdots \\ \vdots & \vdots & \vdots & \ddots & & & & & & 0 \\ 0 & 0 & 0 & \dots & 1 & b_{r\,r+1} & \dots & b_{rn} & & 0 \\ \vdots & \vdots & \vdots & \vdots & & \vdots & & \vdots & & \vdots \\ 0 & 0 & 0 & \dots & & 0 & \dots & 0 & & 0 \end{array}\right] r \quad m$$

Solving for the unknown corresponding to the leading 1's, we have

$$x_1 = -b_{1r+1}\, x_{r+1} - b_{1r+2} + x_{r+2} - \ \dots - b_{1n}x_n$$

$$x_2 = -b_{2r+1}\, x_{r+1} - b_{2r+2} + x_{r+2} - \ \dots - b_{2n}x_n$$

$$\vdots$$

$$x_r = -b_{r\,r+1}\, x_{r+1} - b_{r\,r+2} + x_{r+2} - \ \dots - b_{rn}x_n$$

Where $x_{r+1}, x_{r+2}, \dots, x_n$ can be assigned arbitrary real values $s_j, j = 1,2,\dots, p, and \ \ p = n - r.$

Thus

$$x = \begin{bmatrix} x_1 \\ x_2 \\ \vdots \\ x_r \\ x_{r+1} \\ x_{r+2} \\ \vdots \\ x_n \end{bmatrix} = \begin{bmatrix} -b_{1r+1}\, s_1 - b_{1r+2}s_2 - \ldots - b_{1n}s_p \\ -b_{1r+2}\, s_2 - b_{2r+2}s_2 - \ldots - b_{2n}s_p \\ \vdots \\ -b_{r\,r+1}\, s_1 - b_{r\,r+2}s_2 - \ldots - b_{rn}s_p \\ s_1 \\ s_2 \\ \vdots \\ s_p \end{bmatrix}$$

$$= s_1 \begin{bmatrix} -b_{1r+1} \\ -b_{2r+1} \\ \vdots \\ -b_{1r+1} \\ 1 \\ 0 \\ 0 \\ \vdots \\ 0 \\ 0 \end{bmatrix} + s_2 \begin{bmatrix} -b_{1r+2} \\ -b_{2r+2} \\ \vdots \\ -b_{r\,r+2} \\ 0 \\ 1 \\ 0 \\ \vdots \\ 0 \\ 0 \end{bmatrix} + \ldots + s_p \begin{bmatrix} -b_{1n} \\ -b_{2n} \\ \vdots \\ -b_{rn} \\ 0 \\ 0 \\ 0 \\ \vdots \\ 0 \\ 1 \end{bmatrix}$$

Since $s_1, s_2, \ldots, s_p$ can be assigned arbitrary real values, we major the following choices for these values:

$$s_1 = 1. \quad s_2 = 0, \ \ldots, \quad s_p = 0$$

$$s_1 = 1. \quad s_2 = 0, \ \ldots, \quad s_p = 0$$

$$\vdots$$

$$s_1 = 1. \quad s_2 = 0, \ \ldots, \quad s_p = 0$$

Obtaining

$$x_1 = \begin{bmatrix} -b_{1r+1} \\ -b_{2r+1} \\ \vdots \\ -b_{1r+1} \\ 1 \\ 0 \\ 0 \\ \vdots \\ 0 \\ 0 \end{bmatrix} + x_2 \begin{bmatrix} -b_{1r+2} \\ -b_{2r+2} \\ \vdots \\ -b_{r\,r+2} \\ 0 \\ 1 \\ 0 \\ \vdots \\ 0 \\ 0 \end{bmatrix} \ldots + x_p = \begin{bmatrix} -b_{1n} \\ -b_{2n} \\ \vdots \\ -b_{rn} \\ 0 \\ 0 \\ 0 \\ \vdots \\ 0 \\ 1 \end{bmatrix}$$

Since

$x = s_2 x_1 + s_2 x_2 + \ldots + s_p x_p.$

We see that $\{x_1, x_2, \ldots, x_p)$ spans the solution space of $Ax = 0/$ Moreover, if we form the equation

$c_1 x_1 + c_2 x_2 + \ldots + c_p x_p = 0,$

Its coefficient matrix is the matrix whose columns are $x_1, x_2, \ldots, x_p.$ if we look at rows $r+1, r+2, \ldots, n$ of this matrix, we readily see that

$c_1 = c_2 = \ldots = c_p = 0.$

Hence $\left\{x_1, x_2 \ldots\ldots, x_p\right\}$ is linearly independent and forms a basis for the solution space of $Ax = 0,$ the null space $of\ A.$

The procedure for finding a basis for the solution space of a homogeneous system $Ax = 0$. Or the null space of A is $m \times n$, is as follow

Step 1. Solve the given homogeneous system by Gauss- Jordan reduction. If the solution contains no arbitrary constant, then the solution space is {0}, which has no basis; the dimension of the solution space is zero.

Step 2. If the solution x contains arbitrary constants, write as a linear combination of vectors

$x_1, x_2, \ldots\ldots, x_p\ with\ s_1, s_2, \ldots\ldots s_p$

Step 3. The set of vectors $\left\{x_1, x_2, \ldots, x_p\right\}$ is a basis for the solution space of $Ax = 0;$ the dimension of the solution space is p.

Remark. in step 1, suppose that the matrix in reduced row echelon form to which $[A \vdots 0]$ has been transformed has r nonzero rows (also r leading $1's$). The $p = n - r$. That is, the dimension of the solution space is $n - r$. Moreover, a solution $x\, to\ Ax = 0\, has\, n - r$ arbitrary constants. $if\ A\, is\, an\ m \times n$ matrix, we refer to the dimension of thr null space of A as the nullity of $A,$ denoted by nullity $A.$

Example

Find a basis for and the given system by the Gauss-Jordan reduction method, we transform the augmented matrix to reduced row echelon form, obtaining (verify)

$$\begin{bmatrix} 1 & 1 & 4 & 1 & 1 \\ 0 & 1 & 2 & 0 & 1 \\ 0 & 0 & 0 & 1 & 2 \\ 1 & -1 & 0 & 0 & 2 \\ 2 & 1 & 6 & 0 & 1 \end{bmatrix} \begin{bmatrix} x_1 \\ x_2 \\ x_3 \\ x_4 \\ x_5 \end{bmatrix} = \begin{bmatrix} 0 \\ 0 \\ 0 \\ 0 \\ 0 \end{bmatrix}.$$

Solution. ***Step 1*** To solve the given system by the Gauss-Jordan reduction method, we transform the augmented matrix to reduce row echelon form, obtaining (verify)

$$\left[\begin{array}{ccccc|c} 1 & 0 & 2 & 0 & 1 & 0 \\ 0 & 1 & 2 & 0 & -1 & 0 \\ 0 & 0 & 0 & 1 & 2 & 0 \\ 1 & -1 & 0 & 0 & 2 & 0 \\ 2 & 1 & 6 & 0 & 1 & 0 \end{array}\right]$$

Every solution is of the form (verify)

Every solution is of the form (verify)

$$x = \begin{bmatrix} -2s-t \\ -2s-t \\ s \\ -2t \\ t \end{bmatrix}$$

Where s and t are real numbers.

Step 2. Every vector in W is a solution and is them of the form given by Equation. We can then write every vector in W as

$$x = \begin{bmatrix} -2 \\ -2 \\ 1 \\ 0 \\ 0 \end{bmatrix} + t \begin{bmatrix} -1 \\ 1 \\ 0 \\ -2 \\ 1 \end{bmatrix}$$

Since s and t can take on any values, we first let $s = 1, t = 0$, and then let $s = 0, t = 1$, obtaining as solutions

$$x_1 = \begin{bmatrix} -2 \\ -2 \\ 1 \\ 0 \\ 0 \end{bmatrix} \quad and\ x_2 = \begin{bmatrix} -1 \\ 1 \\ 0 \\ -2 \\ 1 \end{bmatrix}$$

Step 3. The set x $\{x_1, x_2\}$ is a basis for W. Moreover, dim $w = 2$

The following example illustrate a type of problems that we will be solving frequently.

Example. Consider he linear system

$$\begin{bmatrix} 1 & 2 & -2 \\ 2 & 4 & -6 \\ 3 & 6 & -9 \end{bmatrix} \begin{bmatrix} x_1 \\ x_2 \\ x_3 \end{bmatrix} = \begin{bmatrix} 2 \\ 4 \\ 6 \end{bmatrix}$$

The set of all solutions to this linear system consist of all vectors of the form

$$x = \begin{bmatrix} 2-2r+3s \\ r \\ s \end{bmatrix}$$

(Verify), which can be written as

$$x = \begin{bmatrix} 2 \\ 0 \\ 0 \end{bmatrix} + r \begin{bmatrix} -2 \\ 1 \\ 0 \end{bmatrix} + s \begin{bmatrix} 3 \\ 0 \\ 1 \end{bmatrix}$$

The set of all solutions to the associated homogeneous system is the two-dimensional subspace of R^3 consisting of all vectors of the form

$$x = r \begin{bmatrix} -2 \\ 1 \\ 0 \end{bmatrix} + s \begin{bmatrix} 3 \\ 0 \\ 1 \end{bmatrix}$$

This subspace is a plane $Đ_1$ passing the origin; the set of al solution to the given nonhomogeneous system is a plane $Đ_2$ that does not pass through the origin and is obtained by shifting $Đ_1$ parallel to itself. Thus, situation is illustrated

The following result, which is important in the study of differential equations, was presented in and its proof left to Exercise of that section.

If x_p is a particular solution to the nonhomogeneous system $Ax = b$, $b \neq 0$, and x_h is a solution to the associated homogeneous system $Ax = 0$, then $x_p + x_h$ is a solution x to the nonhomogeneous linear system $Ax = b$ can be written as $x_p + x_h$, where x_p is a particular solution to the given nonhomogeneous system and x_h is a solution to the associated homogeneous system $Ax = 0$.

THE RANK OF A MATRIX AND APPLICATIONS

In this section we obtain another effective method for finding a basis for a vector space V spanned by a given set of vector $S = \{v_1, v_2 \ldots.., v_n\}$. in this proof of .we developed a technique for choosing a basis for V that is a subset of S. The method to be developed in this section produces a basis for V that is non-guaranteed to be a subset of S. We shall also attach a unique number to a matrix A that we letter show gives us information about the dimension of the solution space of a homogeneous system eith coefficient matrix A.

Definition. Let

$$A = \begin{bmatrix} a_{11} & a_{12} & \ldots & a_{1n} \\ a_{21} & a_{22} & \ldots & a_{2n} \\ & \vdots & & \\ a_{m1} & am_2 & \ldots & a_{mn} \end{bmatrix}$$

Be an $m \times n$ matrix. The row of A,

$$v_1 = (a_{11}, a_{12}, \ldots\ldots, a_{1n})$$

$$v_2 = (a_{21}, a_{22}, \ldots\ldots, a_{2n})$$

$$\vdots$$

$v_m = \left(am_1, am_2 ,\ldots\ldots, a_{mn}\right)$

Considered as vector in R^n span a subspace of R^n. called the row space of A. similarly the columns of A.

$$w_1 = \begin{bmatrix} a_{11} \\ a_{21} \\ \vdots \\ a_{m1} \end{bmatrix}, \quad w_1 = \begin{bmatrix} a_{12} \\ a_{22} \\ \vdots \\ a_{m2} \end{bmatrix}, \ldots\ldots, \quad w_n = \begin{bmatrix} a_{1n} \\ a_{2n} \\ \vdots \\ a_{mn} \end{bmatrix}.$$

Considered as vector in R^m span a subspace of R^m called the column space of A.

Theorem. $if\ A\, and\ B\, are two\ m \times n$ row equivalent matrices, then the row spaces of A and B are equal.

Proof. If A and B are row equivalent, then the rows of B are obtained from those of A by a finite number of the three elementary row operations. Thus, each row of B is a linear combination of the rows of A. Hence the row space of B by a finite number of the three elementary row operations, so the row space of A is contained in the row space of B. Hence the row spaces of A and B are equal.

It follow from that if we take a given matrix A and transform it to reduced row echelon from B, then the row spaces of A and B are equal. Furthermore, recall from exercise in that the nonzero rows of a matrix that is in reduced row echelon form are linearly independent and thus form a basis for its row space. We can use this method to find a basis for a vector space spanned by a given set of vectors in R^n, as illustrated on row spaces of A and B are equal. Furthermore, recall from exercise in that the nonzero rows of a matrix that is in reduced row echelon form are linearly independent and thus form a basis for its row space. We can use this method to find a basis for a vector space spanned by a given set of vectors in R^n, as illustrated on

Example. Let $S = \{v_1, v_2, v_3, v_4\}$, where

$v_1 = (1, -2, 0, 3, -4)$, $\qquad v_2 = (3, 2, 8, 1, 4)$,

$v_3 = (2,3,7,2,3)$ $\qquad v_4 = (-1,2,0,4,-3)$,

And let V be the subspace of R^m given by $V = span$ S. Find a basis for V.

Solution. Note that V is the row spaces of the matrix A whose rows are the given vectors

$$A = \begin{bmatrix} 1 & -2 & 0 & 3 & -4 \\ 3 & 2 & 8 & 1 & 4 \\ 2 & 3 & 7 & 2 & 3 \\ -1 & 2 & 0 & 4 & -3 \end{bmatrix}.$$

Using elementary row operations, we find that A is row eqvalent to the matrix (verify)

$$B = \begin{bmatrix} 1 & 0 & 2 & 0 & 1 \\ 0 & 1 & 1 & 0 & 1 \\ 0 & 0 & 0 & 1 & -1 \\ 0 & 0 & 0 & 0 & 0 \end{bmatrix},$$

Which is in reduced row echelon form. The row spaces of A and B are identical and a basis for the row space of B consist of the nonzero row of B.*Hence*

$$w_1 = (1,0,2,0,1), \quad w_2 = (0,1,1,0,1), \quad and \quad w_3 = (0,0,0,1,-1)$$

Form a basis for V.

We may now summarize the method used in example to find a basis for the subspace V of R^n given by $V = span\, S$, where S is a set of vector in R^n.

The procedure for finding a basis for the subspace $V\ of\ R^n\ given\ by V = span\, S$ where $S = \{v_1, v_2, \ldots\ldots, v_n\}$ is a set of vectors in R^n that are given in row form, is as follows

Step 1. Form the matrix

$$A = \begin{bmatrix} v_1 \\ v_2 \\ \vdots \\ v_k \end{bmatrix}$$

Whose rows are the given vectors in S.

Step 2. Transform A to reduced row echelon form, obtaining the matrix B.

Step 3. The nonzero rows of B form a basis for V.

f course, the basis that has been obtained by the procedure used in produced a basis that is not a subset of the spanning set. However, the basis for subspace $V\ of\ R^n$ that is obtained by the method of where the leading 1's occur in columns $j_1, j_2, \dots\dots j_k$, then it can be shown that $v = a_{ij}v_1 + aj_2 v_2 + \dots\dots + aj_k v_k$

Example. Let V be the subspace of example. Given that the vector $v = (5, 4, 14, 6, 3)$

Is in V, write V as a linear combination of the basis determined in

Solution. We have $j_i = 1,\ j_2 = 2, and\ \ j_3 = 4,$ so $v = 5w_1 + 4w_2 + 6w_3$.

Remark. The solution also yields a basis for the row space of matrix A in that example. Observer that the vectors in this basis are not rows of matrix A.

Definition. The dimension of the row space of A is called the row rank of A, and then dimension of the column space of A is called the column rank of A.

Example. Find a basis for the row space of the matrix A defined in the solution that contains only row vectors from A. Also compute the row of the matrix A.

Using the procedure in the alternative proof of, we form the equation

$$c_1(1,-2,0,3,-4)+c_2(3,2,8,1,4)+c_3(2,3,7,2,3)+c_4(-1,2,0,4,-3)=(0,0,0,0,0),$$

whose augmented matrix is

$$\left[\begin{array}{cccc|c} 1 & 3 & 2 & -1 & 0 \\ -2 & 2 & 3 & 2 & 0 \\ 0 & 8 & 7 & 0 & 0 \\ 3 & 1 & 2 & 4 & 0 \\ -4 & 4 & 3 & -3 & 0 \end{array}\right] = \left[A^T \vdots 0\right];$$

That is the coefficient matrix is A^T. transforming the augmented matrix $\left[A^T \vdots 0\right]$; in to reduced row echelon form, we obtain (verify)

$$\left[\begin{array}{cccc|c} 1 & 3 & 2 & \frac{11}{24} & 0 \\ 0 & 1 & 0 & -\frac{49}{24} & 0 \\ 0 & 0 & 1 & \frac{7}{3} & 0 \\ 0 & 0 & 0 & 0 & 0 \\ -4 & 4 & 3 & -3 & 0 \end{array}\right] = \left[A^T \vdots 0\right];$$

Since the leading 1's in occur in columns 1,2, and 3, we conclude that the first three rows of A form a basis for the space of A. That is,

$$\{(1,-2,0,3,-4),(3,2,8,1,4),(2,3,7,2,3)\}$$

is a basis for the row space A. The row rank of A is 3.

Example: Find a basis for the column space of the matrix A defined in the solution of example and compute the column rank of A,

Solution Writing the columns of A as row vectors, we obtain the matrix A^T, which when transformed to reduced row echelon form is

$$\begin{bmatrix} 1 & 0 & 0 & \frac{11}{24} \\ 0 & 1 & 0 & -\frac{49}{24} \\ 0 & 0 & 1 & \frac{7}{3} \\ 0 & 0 & 0 & 0 \\ 0 & 0 & 0 & 0 \end{bmatrix}$$

The vectors $\left(1,0,0,\frac{11}{24}\right),\left(0,1,0-\frac{49}{24}\right),and\left(0,0,1,\frac{7}{3}\right)$ from a basis for the row space $of\ A^T$ Hence the vectors

$$\begin{bmatrix}1\\0\\0\\\frac{11}{24}\end{bmatrix},\begin{bmatrix}0\\1\\0\\-\frac{49}{24}\end{bmatrix},\ and\begin{bmatrix}0\\0\\1\\\frac{7}{3}\end{bmatrix}$$

Form a basis for the column space of A and we conclude that the column rank of A is 3.

Solution 2. If we're to find a basis for the column space of A that contain only the column vectors from A we follow the procedure developed in the proof of forming the equation

$$c_1\begin{bmatrix}1\\3\\2\\-1\end{bmatrix}+c_2\begin{bmatrix}-2\\2\\3\\2\end{bmatrix}+c_3\begin{bmatrix}0\\8\\7\\0\end{bmatrix}+c_4\begin{bmatrix}3\\1\\2\\4\end{bmatrix}+c_5\begin{bmatrix}-4\\4\\3\\-3\end{bmatrix}=\begin{bmatrix}0\\0\\0\\0\end{bmatrix}$$

whose augmented matrix is $=[A\vdots 0]$. Transforming this matrix to reduced row echelon form, we obtain

$$\left[\begin{array}{ccccc|c}1&0&2&0&1&0\\0&1&1&0&1&0\\0&0&0&1&-1&0\\0&0&0&0&0&0\end{array}\right]$$

Since the leading 1's occur in columns 1, 2, and 4, we conclude that the first second and fourth column of A from a basis for the column space A. That is

$$\left\{\begin{bmatrix}1\\3\\2\\-1\end{bmatrix}\cdot\begin{bmatrix}-2\\2\\3\\2\end{bmatrix}\cdot\begin{bmatrix}3\\1\\2\\4\end{bmatrix}\right\}$$

Is a basis for the column space of A. The column rank l of $A is 3$.

From we observe that the row and rank and column rank of A are equal. This is always true, and is very important result in linear algebra. We now turn to the proof of this theorem.

Theorem. The row rank and column rank of the mxn matrix $A=\left[a_{ij}\right]$ are equal.

Proof. Let $w_1, w_2, \ldots, w_n$ demote the columns of $A.$ To determine the dimension of the column space of A , we use the procedure in the alternative proof of theorem . Thus, we consider the equation

$c_1 w_1 + c_2 w_2 + \ldots + c_n w_n = 0$

We now transform the augmented matrix. $[A \vdots 0]$, of this homogeneous system to reduce row echelo9n form. The vectors corresponding to the columns containing the leading 1's

But this number is also the number of nonzero rows in the reduce row echelon form matrix that is the row equivalent to A, so it is. The row rank of $A.$ Thus row rank of $A.$ = column rank A.

Definition. Since row rank $A = column\, rank\, A,$ we now merely refer to the rank of an $m \times n$ matrix and write rank $A.$

We next summarize the procedure for computing the rank of a matrix. The procedure for computing the rank of the matrix A is as follows.

Step 1. Using elementary row operation, transform A to a matrix B in reduced row echelon form.

Step 2. Rank $A =$ the number of nonzero rows of $B.$

If A is an mxn matrix, we have defined the nullity of A as the dimension of the null space of $A.$ that is , the dimension of the solution of the solution space of $Ax = 0.$ if A is transformed to a matrix B in reduced row echelon from having r nonzero rows,

Then we know that the dimension of the solution space $Ax = 0\, is\, n - r. Since\, r$ is also the rank of $A.$ we have obtained a fundamental relationship between the rank and nullity of $A.$ which we state in the following theorem

Theorem. $if\ A\ is\ an\ m \times n$ matrix. Then rank $A + nullity\ A = n$

Example. Let

$$A = \begin{bmatrix} 1 & 1 & 4 & 1 & 2 \\ 0 & 1 & 2 & 1 & 1 \\ 1 & -1 & 0 & 0 & 2 \\ 2 & 1 & 6 & 0 & 1 \end{bmatrix}$$

Which was defined when A is transformed to reduce row echelon form we obtain

$$\begin{bmatrix} 1 & 0 & 2 & 0 & 1 \\ 0 & 1 & 2 & -1 & \\ 0 & 0 & 0 & 1 & 2 \\ 0 & 0 & 0 & 0 & 0 \\ 0 & 0 & 0 & 0 & 0 \end{bmatrix}$$

Then rank $A = 3$ and nullity $A = 2.$ This agrees with the result obtained in the solution of, where we found that the dimension of the solution space of $Ax = 0$ is 2

The following example will be used to illustrate geometrically some of the ideas discussed above.

Example. Let

$$A = \begin{bmatrix} 3 & -1 & 2 \\ 2 & 1 & 3 \\ 7 & 1 & 8 \end{bmatrix}$$

Transforming A to reduce row echelon form we obtain (verify)

$$\begin{bmatrix} 1 & 0 & 1 \\ 0 & 1 & 1 \\ 0 & 0 & 0 \end{bmatrix}$$

So, we conclude the following

Rank $A = 2$

Dimension of row space $A = 2,$ so the row space of A is a twp dimensional subspace of R^3, that is a plane passing through the origin

From the reduced row echelon form matrix that A has been transformd to, we see that every solution to the homogeneous system

$Ax = 0 is$ of the form

$$x = \begin{bmatrix} -r \\ -r \\ r \end{bmatrix}$$

Where r is an arbitrary constant (verify), so the solution space of this homogeneous system, or the null space $of\ A.$ is a line passing through the origin. Moreover, the dimension of the null space of $a.$ or the nullity of $a,$ IS 1. Thus, has been verified.

Of course, we already know that the dimension of the column space of A is also a two-dimensional subspace of R^3, that is, a plane passing through the origin. These results are illustrated

The Rank of a square matrix can be used to determine whether the matrix is singular or nonsingular, as the following theorem shows.

Theorem. An nxm matrix A is nonsingular if and only if rank $A = n.$

Proof. Suppose that A is nonsingular. Then A is row equivalent to L_n and so rank $A = n.$

Convulsively, let rank $A = n$. Suppose that A is row equivalent to a matrix B in reduced row echelon form. Then rank $B = n,$ so the rows of A must be linearly independent. Hence B has no zero rows and since it is in reduced row echelon form, it must be I_n. Thus A is row equivalent to I_n and so A is nonsingular

An immediate consequence if is the following corollary which is a criterion for thr rank of an nxn matrix to be n.

Corollary. If $A is an nxn$ matrix, then rank $A =$ n if and only if $\det(A) \neq$.

Proof. Another result easily obtained form is contained in the following corollar7y.

Corollary. $Let\ A be an nxn$ matrix **b** if and only if rank $A = n.$

Proof, The following corollary gives another method of testing whether n gives vector in R^n are linearly dependent or linearly independent.

Corollary. $Let\ S = \{v_1, v_2, \ldots, v_n\}$ be a set of n vectors In R^n and let A be the matrix whose rows (columns) are the vectors in S. Then S is linearly independent.

Proof. For $n > 4$. the method of Corollary for testing linear dependence is not as efficient as the direct method .calling for the solution of a homogeneous system.

Corollary. The homogeneous system $Ax = 0\ of\ n$ linear equationsin n n unknowns have a nontrivial solution if and if rank $A < n$.

Proof. The follows form and from the fact that $Ax = 0$ has a nontrivial solution if and only if A is singular

Example Let

$$A = \begin{bmatrix} 1 & 2 & 0 \\ 0 & 1 & 3 \\ 2 & 1 & 3 \end{bmatrix}$$

If we transform A to reduced row echelon form B, we find that $B = I_3$ (verify). Thus rank $A = 3$ and matrix A is nonsingular. Moreover, the homogeneous system $Ax = 0$ has only the trivial solution.

Example. Let $$A = \begin{bmatrix} 1 & 2 & 0 \\ 0 & 1 & -3 \\ 1 & 3 & 3 \end{bmatrix}$$

Then A is row equivalent to

$$\begin{bmatrix} 1 & 0 & -6 \\ 0 & 1 & 3 \\ 0 & 0 & 0 \end{bmatrix},$$

A matrix in reduced row echelon form. Hence rank $A < 3$, and A is singular. Moreover, $Ax = 0$ has a nontrivial solution.

Application of rank of the linear system $Ax = b$.

In corollary we have seen that the rank of A provides us with information about the exercise of a nontrivial solution to the homogeneous system $Ax = 0.$ We shall now obtain some result that use the rank of A tp provide information about the solutions to the linear system $Ax = b$, wher b is an arbitrary $n \times 1$ matrix/ when $b \neq =$; the linear system is said to be nonhomogeneous .

Theorem. The linear system $Ax = b\, has$ a solution if and only if rank $A = rank\left[A \vdots b\right];$ that is, if and only of the coefficient and augmented matrices are equal.

Proof. First, observe that if $A = \left[a_{ij}\right] is\ m \times n,$ then the given linear system may be

$$x_1 \begin{bmatrix} a_{11} \\ a_{21} \\ \vdots \\ a_{m1} \end{bmatrix} + x_2 \begin{bmatrix} a_{12} \\ a_{22} \\ \vdots \\ a_{m2} \end{bmatrix} + \ldots + x_n \begin{bmatrix} a_{1n} \\ a_{2n} \\ \vdots \\ a_{mn} \end{bmatrix} = \begin{bmatrix} b_1 \\ b_2 \\ \vdots \\ b_m \end{bmatrix}$$

Suppose now that $Ax = b$ has a solution. Then there exist values $of\ x_1, x_2 \ldots\ldots, x_n$ that satisfy the equation. Thus, b is a linear combination of the columns of A and so belong to the columns space $A.$ Hence, rank $A = rank\left[A \vdots b\right]$

Conversely, suppose that rank $A = rank\left[A \vdots b\right]$. Then b is the in the column space of $A,$ which means that we find value $x_1, x_2, \ldots, x_n$ that satisfy. Hence $Ax = b$ has a solution.

Remark. The result given in, while of interest, is not of great computational value, since we usually are interested in finding a solution rather than in knowing whether or not a solution exists.

Example. Consider the linear system

$$\begin{bmatrix} 2 & 1 & 3 \\ 1 & -2 & 2 \\ 0 & 1 & 3 \end{bmatrix} \begin{bmatrix} x_1 \\ x_2 \\ x_3 \end{bmatrix} = \begin{bmatrix} 1 \\ 2 \\ 3 \end{bmatrix}$$

Since $rank\ A = [A \vdots b] = 3.$ the linear system has a solution.

Example. The linear system

$$\begin{bmatrix} 1 & 2 & 3 \\ 1 & -3 & 4 \\ 2 & -1 & 7 \end{bmatrix} \begin{bmatrix} x_1 \\ x_2 \\ x_3 \end{bmatrix} = \begin{bmatrix} 4 \\ 5 \\ 6 \end{bmatrix}$$

Has no solution because rank $A = 2\,and$ rank $[A \vdots b] = 3$ (verify)

We now extended our list of Nonsingular Equivalences.

List of nonsingular Equivalences

The following statements are equivalent for $n\,x\,n$ matrix A.

1. A is nonsingular
2. $Ax = 0$ has only the trivia; solution.
3. A is row equivalent to I_n.
4. The linear system $Ax = b$ has a unique solution for every $n\,x\,1$ matrix b
5. det $(A) \neq 0.$
6. A has rank $n.$
7. A has nullity $0.$
8. The row of A form a linearly independence set n vectors in $R^n \ldots$
9. The columns of A form a linearly independent set n vectors in R^n.

Definition. (Rank of a Matrix). The **rank** of a matrix A is equal to the numbers of nonzero rows in its echelon (or reduced echelon) form or the order of I_r in the canonical form A .

A more formal definition of the rank of a matrix is given in the later section also. Now to find the rank of A we just reduce A to its echelon (or reduce echelon) form or canonical form and count its nonzero rows.

Example 17. Find the rank of the matrix

$$A = \begin{bmatrix} 5 & 9 & 3 \\ -3 & s5 & 6 \\ -1 & -5 & -3 \end{bmatrix}$$

Solution. To find the rank of A, we reduce to an echelon form. Thus

$$A = \tilde{R} \begin{bmatrix} -1 & -5 & -3 \\ -3 & 5 & 6 \\ 5 & 9 & 3 \end{bmatrix} \quad by\, R_{13}$$

$$\tilde{R} = \begin{bmatrix} 1 & 5 & 3 \\ -3 & 5 & 6 \\ 5 & 9 & 3 \end{bmatrix} \quad by(-1)R_1$$

$$\tilde{R} \begin{bmatrix} 1 & 5 & 3 \\ 0 & 20 & 15 \\ 0 & -16 & -12 \end{bmatrix} by\, R_2 + 3R_1 \; and \; R_3 - 5R_1$$

$$\tilde{R} \begin{bmatrix} 1 & 5 & 3 \\ 0 & 1 & 3/4 \\ 0 & -16 & -12 \end{bmatrix} by \frac{1}{20} R_2$$

$$\tilde{R} \begin{bmatrix} 1 & 5 & 3 \\ 0 & 1 & 3/4 \\ 0 & 0 & 0 \end{bmatrix} \quad by\, R_3 + 16R_2$$

This is an echelon form A and the number of its nonzero rows is 2. Hence the rank of $A is 2$.

SUMMARY AND DISCUSSION

In this chapter, we discussed about an important topic of matrices. They play very important role in many real-life problems. The basic concept of matrices is necessary for all fields.

Definition. Let F *denote* the field of real or complex numbers. A rectangular array of m n elements $a_{ij} \in F, (1 \le i \le m, 1 \le j \le n)$ arranged in rows and n columns and enclosed by square brackets such as

$$\begin{bmatrix} a_{11} & a_{12} & \cdots & a_{1j} & \cdots & a_{1n} \\ a_{21} & a_{22} & \cdots & a_{2j} & \cdots & a_{2n} \\ \vdots & \vdots & \cdots & \vdots & \cdots & \vdots \\ a_{i1} & a_{i2} & \cdots & a_{1j} & \cdots & a_{1n} \\ \vdots & \vdots & \cdots & \vdots & \cdots & \vdots \\ a_{m1} & a_{m2} & \cdots & a_{mj} & \cdots & a_{mn} \end{bmatrix}$$

is called an $m \times n$ over the field, F (or with entries from the fields F).

Rectangular Matrix. If a matrix A has m rows and n columns, then A is said to be a rectangular matrix of order (or size) $(m \times n)$. In case $m = n$,

Square Matrix. A is said to be a square matrix of order n. If $A = \left[a_{ij} \right]_{m \times n}$, $[a_{i1} \, a_{i2} \, . \, . \, . \, a_{in}]$ is a $1 \times n$ matrix and is called vector. Similarly, a single column

$$\begin{bmatrix} a_{i1} \\ a_{i2} \\ \vdots \\ a_{mj} \end{bmatrix}$$

is an $m x 1$ matrix and is called column vector.

Upper (Lower) triangular matrix. An upper (lower) triangular matrix (or a triangular matrix) square matrix all of whose elements below (above) the main diagonal (running from upper left to lower right corner) are zero. Thus

$$A = \begin{bmatrix} a_{11} & a_{11} & \cdots & a_{1n} \\ 0 & a_{11} & \cdots & a_{1n} \\ \vdots & \vdots & \cdots & \vdots \\ 0 & 0 & \cdots & a_{nn} \end{bmatrix}$$

is an upper triangular matrix. $a_{11}, a_{22} \ldots, a_{nn}$, are called elements of the main diagonal .

Diagonal Matrix. A square matrix, all of whose elements are zero except those in the main diagonal, is called a diagonal matrix. Thus, a square matrix A = $\left[a_{ij} \right]$ is a diagonal matrix if $a_{ij} = 0$ whenever $i \ne j$.

$$\begin{bmatrix} a & 0 & 0 \\ 0 & b & 0 \\ 0 & 0 & c \end{bmatrix}$$

Scalar Matrix. A square matrix of size n in which each element of the main diagonal is one and the same nonzero number $k \in F$ and all other elements are zero is called a scalar matrix of order n.

Thus, $\begin{bmatrix} k & 0 & 0 & 0 \\ 0 & k & 0 & 0 \\ 0 & 0 & k & 0 \\ 0 & 0 & 0 & k \end{bmatrix}$ is a scalar matrix of order 4.

Every scalar matrix is a diagonal matrix

When each $k = 1$ (the identify element of F with respect to multiplication), the scalar matrix of order n is called the identity matrix and is denoted by $\boldsymbol{I}_n$. Thus

$$\boldsymbol{I}_3 = \begin{bmatrix} 1 & 0 & 0 \\ 0 & 1 & 0 \\ 0 & 0 & 1 \end{bmatrix}, \boldsymbol{I}_4 = \begin{bmatrix} 1 & 0 & 0 & 0 \\ 0 & 1 & 0 & 0 \\ 0 & 0 & 1 & 0 \\ 0 & 0 & 0 & 1 \end{bmatrix}$$

are identify matrices of order 3 and 4 respectively.

ALGEBRA OF MATRICES

Equality of Matrices. Two $m \times n$ matrices $A = [a_{ij}]$ and $B = [b_{ij}]$ are equal if and only if $a_{ij} = b_{ij}$ for each j. When A and B are equal, we write $A = B$.

Zero Matrix. A matrix, whose every element is zero, is called a zero matrix. If it has m rows and n columns, we denote it by $0_{m \times n}$ or simply by 0 if there is no danger of ambiguity about the number of its rows and columns.

Addition of Matrices. Two matrices are said to be conformable for addition when they have the same number of rows and the same number of columns. Thus if $A=\left[a_{ij}\right]$ and $B=\left[b_{ij}\right] are m x n$ matrices over the same field F, then they can be added and their sum is the matrix $A+B=\left[a_{ij}+b_{ij}\right]$ *of* order $m\times n$. *That* is, to find the sum $A+B$ of two matrices A and B of the same order we add their corresponding elements.

Additive Inverse of a Matrix. Given an $m\times n$ matrix $A=\left[a_{ij}\right]$, we define $-A=\left[-a_{i}\right]$. Thus $-A is an m\times n$ *matrix* and by definition $A+(-A)=\left[a_{ij}+\left(-a_{ij}\right)\right]=0=\left[a_{ij}+\left(-a_{ij}\right)\right]=(-A)+A.$ Thus $-A$ is the additive inverse of A. Also, $0+A=A+0, i.e., 0$ is the identity with respect to $+$.

Theorem. For any three $m xn$ matrices A, B and $B and$ C, we have $A+(B+C)=(A+B)+C$ (Associative Law for Matrix Addition)

Subtraction of Matrices. If A and B are two matrices of the same order, we define

$A-B \, as \, A-B=A+(-B).$

Multiplication of a Matrix by a Scalar. Let A + $\left[a_{ij}\right]$ and $k\in F$. Then we define the scalar multiple $k A of A by k as K A=\left[ka_{ij}\right] .i.e., kA$ is the matrix obtained by multiplying each element of $A by k$. The additive inverse of a matrix A can also be defined as $-A=(-1)A$, where 1 is the multiplicative identity of F.

Multiplication of Matrices. Two matrices A and B are said to be conformable for the product AB if the number of columns in A is equal to the number of rows in B. Thus if A is an $m\times n$ matrix and B is an $n\times -p$ matrix, then the product $C=AB$ of the matrices A and B is an $m\times p \, matrix \, C$ whose entries are determined as follows: Let $A=\left[a_{ij}\right]_{m\times m}$ and $B=\left[b_{ij}\right]_{n\times p}$ then $Ab=C, where$ $C=\left[c_{ij}\right]_{m\times p}$ and

The commutative law for multiplication of matrices does not hold in general.

Theorem. If the matrices A, B and C are conformable for the indicated sums and products then

$A(BC) = (AB)C$ (Associate Law for Matrix Multiplication)

$A(B+C) = AC + BC$ (Left Distributive Law for Matrix Multiplication)

$(A+B)C = AC + BC$ (Right Distributive Law for Matrix Multiplication)

$K(AB) = (KA)B = A(kB)$, where $k \in F$.

Transpose of a Matrix. Let $A = [a_{ij}]$ be an $m x n$ matrix interchanging rows columns of A. Thus $A^T = [b_{ij}]$, where $b_{ij} = a_{ij}$. In other words, $(i, j)th$ elemnt of $A^T = (j,i)th$ element of A. Thus the first row of A^T is the first column of A, second row of A^T is the second column of A and so on.

Important properties. If the matrices A and B are conformable for the sum A + B and the product AB, then

$$(A \pm B)^T = A^T \pm B^T$$

$$(A^T)^T = A$$

$$(kA)^T = kA^T, \quad wheer\ k \in F$$

$$(AB)^T = B^T A^T$$

Definition. Let A be a square matrix of order n We define $A^n = A\ .A\ .A \ldots A\ (n\ factors)$

It can be proved by induction that, for all $m, n \in N$

$$A^m . A^n = A^{mxn}$$

$$(A^m)^n = A^{mn}$$

We also define $A^0 = l_n$

Also, if A and B are conformable for multiplication then $(AB)^n \neq A^n B^n$ in general However, $(AB)^n = A^n B^n$ holds if AB=BA

Periodic Matrix. A square matrix A for which $A^{k+1} = a, (k\ being\ a\ positive\ integer)$ is called a periodic matrix. If k is the least positive integer for which A^{k+1}=A then A is said to be of period if k if $k = 1$, so that $A^2 = A,\ then\ A$ called an idempotent matrix.

Nilpotent. A square matrix A for which $A^p = 0, (\ p\ being\ a\ positive\ integer)$ is called nilpotent. If is the least positive integer for which A^p=0 then is said to be nilpotent of nil potency index p.

Involuntary matrix. A square matrix A such $that\ A^2 = I$ is called an involuntary matrix.

Definition. A square A for whioch A^T=A is called a symmetric matrix A square matrix A for which A^T=-A is called a skew symmetric matrix.

Note that the product of two symmetric matrices may not be symmetric. That is if

$A^T - A,\ B^T = B$ then $(AB)^T = B^T A^T = BA \neq AB$ is general

Conjugate Matrix. Let A=[a_{ij}] be a matrix with complex entries. Then the matrix [a_{ij}] obtaine from A by replacing each a_{ij} by its complex conjugate, is called the conjugate matrix of A and is denoted by A (read A conjugate). $(A)^T$ is also denoted by A^H.

Hermitian Matrix. A is Hermitian matrix if $A^H = A.$

Skew Hermitian Matrix. A square matrix $A = \left[a_{ij}\right]$ over C for which $(A)^T = -A$ is called skew Hermitian. As for the product of two symmetric matrices, the product of two Hermitian matrices A and B need not be Hermitian, That $(AB)^H = B^H A^H = BA \neq AB\ in\ general$.

Theorem. Let A and B be the square matrices of the same order. Then

(i) $\left(A^H\right)^H = A,$

(ii) $(A \pm B)^H = A^H \pm B^H$

(iii) $(cA)^H = (c)A^H,$

(iv) $(AB)^H = B^H A^H.$

Partition Matrix and Sub Matrices. Some time we partition a matrix into blocks of elements and consider such blocks as elements of the matrix. These blocks are called sub matrices of the original matrix and are labeled in the same manner as elements of a matrix. A matrix can be partitioned into sub matrices in many different ways and the manner in which it is to be partitioned is often indicated by drawing horizontal and vertical lines between selected rows and columns.

Let A ,B be two $m x n$ matrices A *and* B are said to be identically partitioned if the corresponding sub matrices C_{ij} *and* D_{ij} *of* A B *are of* the same order (i.e have the same size).

In such a case $C_{ij} \pm D_{ij}$ are both defined.

Elementary Row Operations. The following operations on a matrix A are called elementary row operations:

(i) interchanging of any two rows $(R_{ij}$ denotes the interchanging of the ith row with the jth row)

(ii) Multiplication of a row of A by any nnonzero real or complex number $(kR_i$ denote that i th row is multiplied by $k \neq 0$).

(iii) Addition of a scalar multiple of one row to another row. $\left(R_j + kR_i\right)$ denotes that k times row i is added to the j th row

Note that each of these row operations on an $m \times n$ matrix can be effected premultiplying A by an $m \times n$ elementary matrix.

Let A be an $m \times n$ matrix. An $m \times n$ matric B is called row equivalent to A *if* B obtained from A by performing a finite sequence of elementary row operations on A. We write $B\tilde{R}A$ to denote B is row equivalent to A.

Echelon Form. An $m \times n$ *matrix* A is said to be in (row) ***echelon*** form (or an echelon matrix)

If it has following properties.

(i) All nonzero rows are above any zero rows (consisting of all zero)

(ii) The first nonzero entry in each nonzero row is to the right of the first nonzero entry of each preceding row. That is, the number of zeros occurring before the first nonzero entry in each nonzero row is greater than the number of zeros that appear before the first nonzero element in any preceding row. In an echelon matrix, the first nonzero entry of a row is called a pivot (or a row leader). There is at the most one pivot in each roe and in each column of an echelon matrix. A column containing a pivot is called a pivot column. An echelon matrix in which each pivot is I and every other entry of the pivot column is zero. Is said to be in row reduced echelon form.

The matrices $\begin{bmatrix} 0 & 5 & 2 & 1 \\ 0 & 0 & 9 & -1 \\ 0 & 0 & 0 & 0 \end{bmatrix}$ and $\begin{bmatrix} 1 & 3 & 0 & 0 \\ 0 & 0 & 1 & 0 \\ 0 & 0 & 0 & 1 \\ 0 & 0 & 0 & 0 \end{bmatrix}$ are in echelon form. The second matrix is in row reduced echelon form. Every matrix is row equivalent to a matrix in echelon form (reduced echelon form).

A square matrix A is nonsingular if an only if it can be written as a product of elementary matrices i.e

$A = E_r E_{r-1} \ldots E_2 E_1$

Elementary Column Operation. The following operations on a matrix A are called elementary column operations.

(i) Interchanging of any two columns of A

(ii) Multiplication of a columns of A by a nonzero scalar.

(iii) Addition of a scalar multiple of one column of A to another column

On applying an elementary column operation to the identify matrix, we obtain a corresponding elementary matrix, which is nonsingular.

Definition. An $m \times n$ matrix B is called column equivalent to an $m \times n$ matrix A if B is obtained by performing a sequence of a finite number of elementary column operations on A. We write $B \tilde{C} A$ to denote B is column equivalent to A.

- An $m \times n$ matrix B is column equivalent to an $m \times n$ matrix A if and only if $B = AQ$, where Q is an $n x n$ elementary matrix.
- If an $m \times n$ matrix B can be obtained from an $m \times n$ matrix A by a finite number of elementary row and column operations, then B is said to be equivalent to A, We write $B \sim A$ to denote B is equivalent to A.
- An $m \times n$ matrix B is equilvalent to an $m \times n$ matrix if A and only if $B = PAQ$ where P and Q are nonsingular and of order $m \times m$ and $n \times n$ respectively.
- Every nonzero $m \times n$ matrix is equivalent to n $m \times n$ matrix D where

$$D = \begin{bmatrix} I & 0 \\ 0 & 0 \end{bmatrix} (D$$

is called the canonical (or normal form) of the matrix A)

Rank of a Matrix. The rank of a matrix A is equal to the numbers of nonzero rows in its echelon (or reduced echelon) form or the order of I_r in the canonical form A.

HOMOGENIOUS SYSTEMS

Homogeneous systems play a central in linear algebra. This will be seen in where the foundations of the subject are all integrated to solve one of the major problems occurring in a wide variety of applications. In this section we deal with several problems involving homogeneous systems that will be fundamental in. Here we are able to focus our attention on these problems without being distracted by the additional material in.

Consider the homogeneous system $Ax = 0.$

- $if\ A\ and\ B\ are\ two\ m \times n$ row equivalent matrices, then the row spaces of A and B are equal.

Definition. The dimension of the row space of A is called the row rank of A, and then dimension of the column space of A is called the column rank of $A.$

- The row rank and column rank of the $m \times n$ matrix $A = \left[a_{ij} \right]$ are equal.
- $if\ A\ is\ an\ m \times n$ matrix. Then rank $A + nullity\ A = n$

The Rank of a square matrix can be used to determine whether the matrix is singular or nonsingular, as the following theorem shows.

- An $n \times m$ matrix A is nonsingular if and only if rank $A = n.$
- If $A\ is\ an\ n \times n$ matrix, then rank $A =$ n if and only if $\det(A) \neq .$
- $Let\ A\ be\ an\ n \times n$ matrix **b** if and only if rank $A = n.$
- $Let\ S = \{v_1, v_2, \ldots, v_n\}$ be a set of n vectors In R^n and let A be the matrix whose rows (columns) are the vectors in $S.$ Then S is linearly independent.
- The homogeneous system $Ax = 0\ of\ n$ linear equationsin n n unknowns has a nontrivial solution if and if rank $A < n.$
- The linear system $Ax = b\,has$ a solution if and only if rank $A = rank\left[A \vdots b\right];$ that is, if and only if the coefficient and augmented matrices are equal.

List of nonsingular Equivalences

The following statements are equivalent for $n \times n$ matrix A.

- A is nonsingular
- $Ax = 0$ has only the trivia; solution.
- A is row equivalent to $I_n.$

- The linear system $Ax = b$ has a unique solution for every $n \times 1$ matrix b
- det $(A) \neq 0.$
- A has rank $n.$
- A has nullity $0.$
- The row of A form a linearly independence set n vectors in R^n...
- The columns of A form a linearly independent set n vectors in R^n.

CHAPTER 2

SYSTEM OF LINEAR EQUATIONS

INTRODUCTION

An equation of the form $a_1x_1 + a_2x_2 + a_3x_3 + \ldots + a_nx_n = b$ is called a linear equation in the variable $x_1, x_2, x_3, \ldots, x_n$. Here $a_1, a_2 \ldots, a_n$ b are constants, usually real numbers A collection of several such equation is called a system of linear equations. The theory of matrices has been usefully employed in various branches of well as applied mathematics. In what follow, we shall apply this theory to the solution of m linear equation in n unknowns.

Consider the m linear equations:

$$\left.\begin{matrix} a_{11}x_1 + a_{12}x_2 + \ldots + a_{1n}x_n = b_1 \\ a_{21}x_1 + a_{22}x_2 + \ldots + a_{2n}x_n = b_2 \\ \vdots \quad \vdots \quad \ldots \quad \vdots \quad \vdots \\ a_{m1}x_1 + a_{m2}x_2 + \ldots + a_{mn}x_n = b_m \end{matrix}\right\}$$

In n unknowns $x_1, x_2, \ldots, x_n,$ where a_{ij}, b_i are scalars, $i = 1, 2, \ldots, m, j = 1, 2, \ldots, n$

Using the matrix notations, the system (1) can be written as

$Ax = b$

Where A, x, b are the matrices as given below:

$$A = \begin{bmatrix} a_{11} & a_{12} \cdots & a_{in} \\ a_{21} & a_{22} \cdots & a_{2n} \\ \vdots & \vdots & \cdots \vdots \\ a_{m1} & a_{m2} \cdots & a_{mn} \end{bmatrix}, \quad x = \begin{bmatrix} x_1 \\ x_2 \\ \vdots \\ x_n \end{bmatrix}, \quad b = \begin{bmatrix} b_1 \\ b_2 \\ \vdots \\ b_m \end{bmatrix}$$

The matrix A is called matrix of coefficients of the system of equations (1), the column of constants b_i forms column vector b and the unknowns x_i from the column vector x.

Definition. The equation of the type given in (1) above with $b \neq 0$, are called system of nonhomogeneous linear equations. If $b = 0$, then the system of equations (1) is known as a homogenous linear system.

Definition. An order $n-tuple\, y = (y_1, y_2, \ldots, y_n)$ is called a solution of the system (1), if, whenever $x_1, x_2 \ldots, x_{nn}$ are replaced by $y_1, y_2, \ldots y_n$ respectively in each equation of (1, the resulting statements are true statements. The vector y is then also said to satisfy the system of equations (1).

For example, the system of linear equations

$$x_1 + 2x_2 = 1$$

$$2x_1 + x_2 = 2$$

has the ordered pair (1,0) as a solution. Here $y = (1, 0)$

Definition. The set S of all ordered $n-tuples\, y = (y_1, y_2, \ldots, y_n$ satisfying the equations (1) is called the solution set of the given equations.

Equivalent of Equations

Definition. Let $Ax = b$,

Where $A = [a_{ij}]_{m \times n}$, $b = \begin{bmatrix} b_1 \\ b_2 \\ \vdots \\ b_m \end{bmatrix}$, $x = \begin{bmatrix} x_1 \\ x_2 \\ \vdots \\ x_n \end{bmatrix}$ (1)

be a given system of linear equation in unknowns. A system of linear equations $A'x = d$

$$A' = \left[a'_{ij}\right]_{m\times n}, \quad d = \begin{bmatrix} b'_1 \\ b'_2 \\ \vdots \\ b'_m \end{bmatrix}, \; x = \begin{bmatrix} x_1 \\ x_2 \\ \vdots \\ x_n \end{bmatrix}$$

Where (2)

is said to be equivalent to the system (1) if matrices

$$\left[A|B\right] = A_b = \left[\begin{array}{cccc|c} a_{11} & a_{12} & \cdots & a_{1n} & b_1 \\ a_{21} & a_{22} & \cdots & a_{2n} & b_2 \\ \vdots & \vdots & \cdots & \vdots & \vdots \\ a_{m1} & a_{m2} & \cdots & a_{mn} & b_m \end{array}\right]$$

and

$$\left[A'|B'\right] = A'_b = \left[\begin{array}{cccc|c} a'_{11} & a'_{12} & \cdots & a'_{1n} & b'_1 \\ a'_{21} & a'_{22} & \cdots & a'_{2n} & b'_2 \\ \vdots & \vdots & \cdots & \vdots & \vdots \\ a'_{m1} & a'_{m2} & \cdots & a'_{mn} & b'_m \end{array}\right]$$

are row equivalent. Since application of an elementary row operation to the matrix implies application of the corresponding operation to the equations (1), we see that the system (1) is equivalent to (2) if (2) is obtained from (1) by any of the following three operations.

Any two equations of (1) are interchanged.

Any equation of (1) is multiplied by a nonzero scalar.

A scalar multiple of an equation of (1) is added to another equation of (1)

The three types of operations listed above are called elementary or a admissible operations for the system (1).

Correspondingly, the two matrices $\left[A|B\right] and \left[A'|B'\right]$, as in (1) and (2) are said to be row equivalent if $\left[A'|B'\right]$ can be obtained from $\left[A|B\right]$ by the following elementary row operations.

Interchange of any two rows of $\left[A|B\right]$,

Multiplying a row of $\left[A|B\right]$ by a nonzero scalar.

Adding a scalar multiple of th row of $\left[A|B\right]$ to its row.\

The matrix

$$\left[A|B\right] = A_b = \left[\begin{array}{cccc|c} a_{11} & a_{12} & \cdots & a_{1n} & b_1 \\ a_{21} & a_{22} & \cdots & a_{2n} & b_2 \\ \vdots & \vdots & \cdots & \vdots & \vdots \\ a_{m1} & a_{m2} & \cdots & a_{mn} & b_m \end{array}\right]$$

is known as the augmented matrix of the system of linear equations (1). It is quite important in the sense that its rank and the rank of (the matrix of coefficients) determine whether the system of equations $\boldsymbol{Ax} = \boldsymbol{b}$ does or does not have a solution.

In fact, the system of equations given by (1) has a unique solution if and only if rank $A = rank\ A_b$. We can use the result of to obtain another method, known as Crammer's rule for solving a linear system of n equations in n unknowns with a nonsingular coefficient matrix.

Let

$$\begin{aligned} a_{11}x_1 + a_{12}x_2 + \ldots + a_{1n}\,x_n &= b_1 \\ a_{21}x_1 + a_{22}x_2 + \ldots + a_{2n}x_n &= b_2 \\ \vdots \qquad \vdots \qquad \vdots \quad \vdots \quad &\quad \vdots \\ a_{n1}x_1 + a_{n2}x_2 + \ldots \quad a_{nn}x_n &= b_n \end{aligned}$$

Gabriel Crammer (1704-1752) was born in Geneva. Switzerland, and lived there all his life. Remaining single, he travelled extensively, taught at the Academic de Calvin, and particle actively in civic affairs.

The rule for solving systems of linear equations appeared in an appendix to his 1750 book, introduction a lines courses algebriques. It was known previously by other mathematicians but was not widely known or very clearly explained until its appearance in Crammer influential work be a linear system of n equations in n unknowns and let $A = [a_{ij}]$ be the coefficient matrix so that we can write the given system as $Ax = b$, where

$$b = \begin{bmatrix} b_1 \\ b_2 \\ \vdots \\ b_n \end{bmatrix}$$

If $\det(A) \neq 0$

$$x_1 = \frac{\det(A_1)}{\det(A)}, \qquad x_2 = \frac{\det(A_1)}{\det(A)} \ldots\ldots x_n = \frac{\det(A_n)}{\det(A)}$$

Where A_1 is the matrix obtained from A by replacing the column of A by .

Proof. If $\det(A) \neq 0$ then, by Theorem is nonsingular. Hence

$$x = \begin{bmatrix} x_1 \\ x_2 \\ \vdots \\ x_n \end{bmatrix} = A^{-1}b = \begin{bmatrix} \frac{A_{11}}{\det(A)} & \frac{A_{21}}{\det(A)} & \cdots & \frac{A_{n1}}{\det(A)} \\ \frac{A_{12}}{\det(A)} & \frac{A_{22}}{\det(A)} & \cdots & \frac{A_{n2}}{\det(A)} \\ \vdots & \vdots & & \vdots \\ \frac{A_{1i}}{\det(A)} & \frac{A_{2i}}{\det(A)} & \cdots & \frac{A_{ni}}{\det(A)} \\ \vdots & \vdots & & \vdots \\ \frac{A_{1n}}{\det(A)} & \frac{A_{2n}}{\det(A)} & \cdots & \frac{A_{nn}}{\det(A)} \end{bmatrix} \begin{bmatrix} b_1 \\ b_2 \\ \vdots \\ b_n \end{bmatrix}$$

This means that

$$x_1 = \frac{A_{1i}}{\det(A)}b_1 + \frac{A_{2i}}{\det(A)}b_2 + \ldots + \frac{A_{ni}}{\det(A)}b_n \, (1 \le i \le n).$$

Now let

$$A_i = \begin{bmatrix} a_{11} & a_1 & \ldots & a_{1i-1} & b_1 & a_{1i+1} & \ldots a_{1n} \\ a_{21} & a_{22} & \ldots & a_{2i-1} & b_2 & a_{2i+1} & \ldots & a_{2n} \\ \vdots & \vdots & & \vdots & \vdots & \vdots & & \vdots \\ a_{n1} & a_{n2} & \ldots & a_{ni-1} & b_n & a_{ni+1} & \ldots & a_{nn} \end{bmatrix}$$

If we evaluate $\det(A_i)$ *by expanding* along the ith column, we find that

$$\det(A_i) = A_{1i}b_1 + A_{2i}b_2 + \ldots + A_{ni}b_n.$$

Hence

$$x_i = \frac{\det(A_i)}{\det(A)}$$

For $i = 1, 2 \ldots, n$. In this expression for x_i, the determination of A_i, det (A_i), cab be calculated by any method. It was only in the derivation of the expression for x_i that we had to evaluate it by expanding along the ith column.

Consider the following linear system

$$-2x_1 + 3x_2 - x_3 = 1$$

$$x_1 + 2x_2 - x_3 = 4$$

$$-2x_1 - x_2 + x_3 = -3.$$

Then

$$|A| = \begin{vmatrix} -2 & 3 & -1 \\ 1 & 2 & -1 \\ -2 & -1 & 1 \end{vmatrix} = -2$$

Hence

$$x_1 = \frac{\begin{vmatrix} 1 & 3 & -1 \\ 4 & 2 & -1 \\ -3 & -1 & 1 \end{vmatrix}}{A} = -\frac{-4}{-2} = 2:$$

$$x_2 = \frac{\begin{vmatrix} 2 & 1 & -1 \\ 1 & 4 & -1 \\ -2 & -3 & 1 \end{vmatrix}}{|A|} = -\frac{-4}{-2} = 2:$$

$$x_2 = \frac{\begin{vmatrix} 2 & 3 & 1 \\ 1 & 2 & 4 \\ -2 & -1 & -3 \end{vmatrix}}{|A|} = -\frac{-8}{-2} = 4:$$

Crammer's rule for solving the linear system $Ax = b,$ where A is $n \times n$. is as follow,

Step 1. Compute det (A). if det $(A)=0.$ Crammer's rule is not application Use Gauss – Jordan reduction.

Step 2. If $\det(A) \neq 0,$ for each i,

$$x_i = \frac{\det(A_i)}{\det(A)}$$

where A_i is the a matrix obtained from A by repacing the ith column A by b.

We note that Cramer 'rule, is, applicable only to the case where we have n equations in n unknowns asnd where the coefficient matrix A is nonsingular. If we have to solve a linear system of n equations in n unknowns whose coefficient matrix is singular, then we must use the Gauss-Jordan reduction method as discussed in. Cramer's rule becomes computationally inefficient for $n > 4,$ and it is better to use the Gauss-Jordan reduction method.

Polynomial Interpolation Revisited

At the end of section we discussed the problem of finding a quadratic polynomial that interpolates the points $(x_1, y_1), (x_2, y_2), (x_3, y_3), x_1 \neq x_2,\ x_1 \neq x_3, and\ x_2 \neq x_3.$ Thus, the polynomial has the form

$$y = a_2 x^2 + a_1 x + a_0$$

This was Equation in section. Substituting the given points in we obtain the linear system

$$a_2x_1^2 + a_1x_1 + a_0 = y_1$$

$$a_2x_2^2 + a_1x_2 + a_0 = y_2$$

$$a_2x_3^2 + a_1x_3 + a_0 = y_{3.}$$

The coefficient matrix of this linear system is

$$\begin{bmatrix} x_1^2 & x_1 & 1 \\ x_2^2 & x_2 & 1 \\ x_3^2 & x_3 & 1 \end{bmatrix}$$

Whose determinant in the Vandermonde determinant, which has the value

$$(x_2 - x_1)(x_3 - x_1)(x_2 - x_3)$$

Since the three given points are districts, the Vandermonde determinants is not zero. Hence, the coefficient matrix of the linear system has a unique solution, thus there is unique interpolating quadratic polynomial. The general proof for n points is similar.

GAUSSIAN ELIMINATION METHOD.

One of the several methods employed to solve a system of n linear equations in n variable, is known as Gaussian Elimination Procedure (named after its discoverer, the German mathematician **C.F. Gauss,** 1777 -1855). Computer programs exist to solve a system of equations by Gaussian Elimination Method and Gauss-Jordan Method described in the following paragraphs. We explain these methods below:

$$\begin{aligned} a_{11}x_1 + a_{12}x_2 + \dots + a_{1n}x_n &= b_1 && (1) \\ a_{21}x_1 + a_{22}x_2 + \dots + a_{2n}x_n &= b_2 && (2) \\ \vdots \quad \vdots \quad \dots \quad \vdots & && \vdots \\ a_{m1}x_1 + a_{m2}x_2 + \dots + a_{mn}x_n &= b_m && (n) \end{aligned}$$

$Ax = b$, where $A = \left[a_{ij}\right]_{n\times n}$,

$$x = \begin{bmatrix} x_1 \\ x_2 \\ \vdots \\ x_n \end{bmatrix}, \; b = \begin{bmatrix} b_1 \\ b_2 \\ \vdots \\ b_n \end{bmatrix}$$

be a system of n equations in n unknowns $x_1, x_2 \ldots, x_n$. we can suppose, without any loss of generality, that coefficient of the first term of the first equation namely a_{11} is not $zero$[1], and has the largest absolute value among the coefficient of the first terms of each equation. Such an element is called pivot and the corresponding equation is known as the pivot equation.

Divide the first equation by a_{11} (the pivot) and obtain

$$x_1 + \frac{a_{12}}{a_{11}} x_2 + \ldots + \frac{a_{1n}}{a_{11}} x_n = \frac{b_1}{a_{11}}$$

Multiply (1) by $a_{i1}, i = 2, 3 \ldots, n$ and subtract from the equations $(2), (3), \ldots, (n)$ respectively, getting a new system of equivalent equations:

$$x_1 + \frac{a_{12}}{a_{11}} x_2 + \ldots + \frac{a_{1n}}{a_{11}} x_n = \frac{b_1}{a_{11}}$$

$$\left(a_{22} - a_{21}\frac{a_{12}}{a_{11}}\right) x_2 + \ldots + \left(a_{2n} - a_{21}\frac{a_{1n}}{a_{11}}\right) x_n = b_2 - a_{21}\frac{b_1}{a_{11}}$$

or $x_1 + a'_{12} x_2 + \ldots + a'_{1n} x_n = b'_1$ (1')

$a'_{22} x_2 + \ldots + a'_{2n} x_n = b'_1$ (2')

$\vdots \quad \ldots \quad \vdots \quad \vdots$

$a'_{n2} x_2 + \ldots + a'_{nn} x_n = b'_n$

Again, we can suppose that at least one of the coefficient $a'_{i2}\, i = 2, \ldots, n$ is different from zero. Since we can rearrange these $(n-1)$ equations, it can be assumed that $a'_{22} \neq 0$ and is a pivot. Carrying out the division of (2') by a'_{22} and then eliminating x_2 from the remaining $n-2$ equations, the new system takes the form

1. For if $a_{11} = 0$, this equation could be interchanged with one in which the first coefficient is nonzero. Of course, we assume that not all

$a_{ij} = 0, i = 1, 2, \ldots, n$. For otherwise, the number of unknowns reduce to $n-1$.

The choice of pivot element is necessary in case where the coefficient are very small. This is because division by small number may increase the coefficients are very small. This is because division by small numbers may increase the coefficients leading to convergence and significant error problems. In our case, we need only 1 in the (1, 1) position and then make all other entries in the first column as zero.

$$
\begin{aligned}
x_1 + a'_{12}x_2 + a'_{13}x_n + \ldots + a'_{1n}x_n &= b'_1 \\
x_2 + a''_{23}x_3 + \ldots + a''_{2n}x_n &= b''_2 \\
a''_{33}x_3 + \ldots + a''_{3n}x_n &= b''_3 \\
\vdots \quad \ldots \quad \vdots \quad & \quad \vdots \\
a''_{n3}x_3 + \ldots + a''_{nn}x_n &= b''_n
\end{aligned}
$$

Continuing this process and eliminating, step by step the variable $x_3, x_4, \ldots, x_{n-1}$ from the remaining $(n-3), (n-4), \ldots, 2$ equations the final form of system of equations becomes

$$
\left.
\begin{aligned}
x_1 + p_{12}x_2 + p_{14}x_3 + \ldots + P_{1n}x_n &= q_1 \\
x_2 + p_{23}x_3 + \ldots + p_{2n}x_n &= q_2 \\
x_{n-1} + P_{(n-1)n}\, x_n &= q_{n-1} \\
x_n &= q_n
\end{aligned}
\right\}
$$

A system of equations of the form (II) is said to be in echelon from. The system (II) is equivalent to the given system and has the same solution set. But in the system (II), x_n is explicitly known's and putting its value in the last but one equation, we obtain x_{n-1} *The value of* x_{n-1} and are then substituted into the preceding equation giving x_{n-2} and, continuing in this way, we find values all the unknowns.

The process of finding the values of the unknowns from the system of equations (II) described above, is called the process of backward substitution or solving the equations recursively.

Observe that the system of equations (II) can be written in the matrix form as:

$$\begin{bmatrix} 1 & p_{12} & p_{14} & \cdots & p_{1n} \\ 0 & 1 & P_{23} & \cdots & p_{2n} \\ 0 & 0 & 1 & \cdots & p_{3n} \\ \vdots & \vdots & \vdots & \cdots & \vdots \\ 0 & 0 & 0 & \cdots & 1 \end{bmatrix} \begin{bmatrix} x_1 \\ x_2 \\ x_3 \\ \vdots \\ x_n \end{bmatrix} = \begin{bmatrix} q_1 \\ q_2 \\ q_3 \\ \vdots \\ q_n \end{bmatrix}$$

i.e $\quad Px = q$

where P is the echelon (upper triangular) matrix obtained from the matrix A by elementary row operations. If E is the product of elementary matrices corresponding to these row operations then

$P = EA \quad and \quad q = Eb$

The given system reduce to (II) if and only if the rank of A is n. If the rank of A is less than n, Gaussian elimination method fails for one following two reasons:

Either we are not able to determine the values of all the variables, or

The apparent inconsistency that all the p_{ij} in a certain row are zero while the corresponding q_i is not zero

Example1. Solve the system of equations:

$x_1 - x_2 + 2x_3 = 0 \qquad (1)$

$4x_1 - x_2 + 2x_3 = 1 \qquad (2)$

$x_1 - x_2 + x_3 = -1 \qquad (3)$

Solution. (First Method). The coefficient of x_1 in the (1, 1) position is already 1. Following the procedure described above, multiplying equation (1) by 4 and subtracting the resulting equation form (2) and subtracting equation (1) from equation (3) respectively, we have

$x_1 - x_2 + 2x_3 = 0 \qquad (1)$

$5x_2 + 6x_3 = 1 \qquad (1)$

$2x_2 - x_3 = -1 \qquad (3)$

Multiplying equation (5) by (2) and subtract from equation (4). The set of equations becomes

$x_1 - x_2 + 2x_3 = 0$ (1)

$x_2 - 4x_3 = 3$ (4)

$2x_2 - x_3 = -1$ (5)

Multiplying equation (6) by (2) and subtract from equation (5). We have

$x_1 - x_2 + 2x_3 = 0$ (1)

$x_2 - 4x_3 = 3$ (6)

$7x_3 = -7$ (7)

Now (7)

$x_2 = -1, \quad x_1 = 1.$

Second Method. (By Gaussian Method)

Here

$$A = \begin{bmatrix} 1 & -1 & 2 \\ 4 & 1 & 2 \\ 1 & 1 & 1 \end{bmatrix}, \quad A_b = \left[\begin{array}{ccc|c} 1 & -1 & 2 & 0 \\ 4 & 1 & 2 & 1 \\ 1 & 1 & 1 & -1 \end{array}\right]$$

So, we apply row operations on A_b to reduce it to the upper triangular form. Thus

$$A_b = \left[\begin{array}{ccc|c} 1 & -1 & 2 & 0 \\ 4 & 1 & 2 & 1 \\ 1 & 1 & 1 & -1 \end{array}\right] \tilde{R} \left[\begin{array}{ccc|c} 1 & -1 & 2 & 0 \\ 0 & 5 & -6 & 1 \\ 1 & 2 & -1 & -1 \end{array}\right] by\, R_2 - 4R_1 \, and \, R_3 - R_1$$

$$\tilde{R} \left[\begin{array}{ccc|c} 1 & -1 & 2 & 0 \\ 0 & 1 & -4 & 3 \\ 1 & 2 & -1 & -1 \end{array}\right] by\, R_2 - 2R_3$$

$$\tilde{R} \left[\begin{array}{ccc|c} 1 & -1 & 2 & 0 \\ 0 & 1 & -4 & 3 \\ 0 & 0 & 7 & -1 \end{array}\right] by\, R_3 - 2R_2$$

$$\tilde{R} \left[\begin{array}{ccc|c} 1 & -1 & 2 & 0 \\ 0 & 1 & -4 & 3 \\ 0 & 0 & 1 & -1 \end{array}\right] by\, \frac{1}{7} R_3$$

So x_3 =-1. By backward substitution, the equations

$x_2 - 4x_3 = 3 \quad and\, x_1 - x_2 + 2x_3 = 0 \ yeild$

$x_1 = 1, \quad x_2 = -1\,,\ x_3 = -1$

Example 2. Solve the following systems of equations by Gaussians elimination method:

$x_1 + 5x_2 + 2x_3 = 9$

$x_1 + x_2 + 7x_3 = 6$

$-3x_2 + 4x_3 = -2$

Solution. Here we apply now row operations on A_b indicated as under. Thus

$$A_b = \left[\begin{array}{ccc|c} 1 & 5 & 2 & 9 \\ 0 & 1 & 7 & 6 \\ 0 & -3 & 4 & -2 \end{array}\right] \tilde{R} \left[\begin{array}{ccc|c} 1 & 5 & 2 & 9 \\ 0 & -4 & 5 & -3 \\ 0 & -3 & 4 & -2 \end{array}\right] by\ R_2 - R_1$$

$$\tilde{R} \left[\begin{array}{ccc|c} 1 & 5 & 2 & 9 \\ 0 & -1 & 1 & -1 \\ 0 & -3 & 4 & -2 \end{array}\right] by\ R_2 - R_3$$

$$\tilde{R} \left[\begin{array}{ccc|c} 1 & 5 & 2 & 9 \\ 0 & -1 & 1 & -1 \\ 0 & 0 & 1 & -1 \end{array}\right] by\ R_3 - 3R_2$$

So, $x_1 + 5x_2 + 2x_3 = 9$

$-x_2 + x_3 = -1$

$x_3 = 1$

By backward substitution, we obtain

$x_3 = 1, \quad x_2 = 2 \quad x_1 = -3.$

Example 3. A yarn merchant sells brands A, B, C of yarn each of which is a blend of Pakistani, Egyptian and American cotton e $x_1\ , x_2\ and\ x_3$ ruoes respectively. Then the above mentioned problem is formulated into the following mathematical model involving equations:

$\frac{1}{4}x_1 + \frac{2}{4}x_2 + \frac{1}{4}x_3 = 40$ (for brand A)

$\frac{2}{4}x_1 + \frac{1}{4}x_2 + \frac{1}{4}x_3 = 50$ (for brand B)

$\frac{2}{4}x_1 \qquad +2x_3 = 240$

So

$$A_b = \left[\begin{array}{ccc|c} 1 & 2 & 1 & 160 \\ 2 & 1 & 1 & 200 \\ 2 & 0 & 3 & 240 \end{array}\right] \tilde{R} \left[\begin{array}{ccc|c} 1 & 2 & 1 & 160 \\ 0 & -3 & -1 & -120 \\ 0 & -4 & 0 & -80 \end{array}\right] \quad By\ R_2 0 - 2R_1\ and\ R_3 - 2R_1$$

$$\tilde{R} \left[\begin{array}{ccc|c} 1 & 2 & 1 & 160 \\ 0 & -4 & 0 & -80 \\ 0 & -3 & -1 & -120 \end{array}\right] \quad by\ R_{23}$$

$$\tilde{R} \left[\begin{array}{ccc|c} 1 & 2 & 1 & 160 \\ 0 & 1 & 0 & 20 \\ 0 & -3 & -1 & -120 \end{array}\right] \quad by - \frac{1}{4}R_2$$

$$\tilde{R} \left[\begin{array}{ccc|c} 1 & 2 & 1 & 160 \\ 0 & 1 & 0 & 20 \\ 0 & 0 & -1 & 60 \end{array}\right] \quad by\ R_3 + 3R_2.$$

Thus, given system is equivalent to

$x_1 + 2x_2 + x_3 = 160$

$x_2 = 20$

$x_3 = 60$

Hence, by backward substitution, we obtain,

$x_1 = RS\,60, \quad x_2 = Rs\,20, \quad x_3 = Rs\,60.$

GAUSS-JORDAN ELIMINATION METHOD

While reducing the given system of linear equation in to the equivalent system (II), we eliminated $x_k, k = 1, 2, \ldots, n-k$ from the $(K+1)th, (K+2)th \ldots nth$ equations only, If we remove x_k from the first $(k-1)$ equations also then x_k will appear only in kth eqution. With this modification we get another equivalent system from the given system expressing each x_k explicitly and we no more require any backward substitution of these for finding the values of other unknowns, This process is called the Gauss-Jordan Elimination Method. The following example explains the working of this method.

Example 4. Find the solution of the following system of linear equation by Gauss-Jordan Elimination Method.

$$2x_1 - x_2 - x_3 = 4$$

$$3x_1 + 4x_2 - 2x_3 = 11$$

$$3x_1 - 2x_2 + 4x_2 = 11.$$

Solution. We reduce the augmented matrix A_b into the reduced echelon as follows:

$$A_b = \begin{bmatrix} 2 & -1 & -1 & 4 \\ 3 & 4 & -2 & 11 \\ 3 & -2 & 4 & 11 \end{bmatrix} \tilde{R} \begin{bmatrix} 3 & 4 & -2 & 11 \\ 2 & -1 & -1 & 4 \\ 3 & -2 & 4 & 11 \end{bmatrix} \quad by\, R_{12}$$

$$\tilde{R} \begin{bmatrix} 1 & 5 & -1 & 7 \\ 2 & -1 & -1 & 4 \\ 3 & -2 & 4 & 11 \end{bmatrix} \quad by\, R_1 - R_2$$

$$\tilde{R} \begin{bmatrix} 1 & 5 & -1 & 7 \\ 0 & -11 & 1 & -10 \\ 3 & -6 & 6 & 0 \end{bmatrix} \quad by\, R_3 - R_2$$

$$\tilde{R} \begin{bmatrix} 1 & 5 & -1 & 7 \\ 0 & -11 & 1 & -10 \\ 3 & -1 & 1 & 0 \end{bmatrix} \quad by \frac{1}{6} R_3$$

$$\tilde{R} \begin{bmatrix} 1 & 5 & -1 & 7 \\ 0 & -10 & 0 & -10 \\ 0 & -1 & 1 & 0 \end{bmatrix} \quad by\, R_2 - R_3$$

$$\tilde{R} \begin{bmatrix} 1 & 5 & -1 & 7 \\ 0 & 1 & 0 & 1 \\ 0 & -1 & 1 & 0 \end{bmatrix} \quad by - \frac{1}{10} R_2$$

$$\tilde{R} \begin{bmatrix} 1 & 5 & -1 & 7 \\ 0 & 1 & 0 & 1 \\ 0 & 0 & 1 & 1 \end{bmatrix} \quad by\, R_3 - R_2$$

$$\tilde{R} \begin{bmatrix} 1 & 5 & 0 & 8 \\ 0 & 1 & 0 & 1 \\ 0 & 0 & 1 & 1 \end{bmatrix} \quad by\, R_1 - R_3$$

$$\tilde{R} \begin{bmatrix} 1 & 0 & 0 & 3 \\ 0 & 1 & 0 & 1 \\ 0 & 0 & 1 & 1 \end{bmatrix} \quad by\, R_1 - 5R_2$$

So $x_1 = 3, \quad x_2 = 1, \quad x_3 = 1.$

Example 5

The Housing Department of the Government plans to undertake four housing projects and list material requirements for the houses in each of the projects as follows:

	Project 1	Project 2	Project 3	Project 4
Paint (in 100 gallons)	1	2	1	1.5
Wood (*in* 10,000 *cu. Ft*)	3	4	2.5	2.5
Brick (in millions)	1	2	1.5	1
Labour (in 1000 hours)	10	10	9	8

If the supplier delivers, 6,800 gallons of paint, 14,200,200 cubic feet of wood, 64 million bricks and 448,000 hours of labor, find the number of houses built for each project.

Solution. Let x_1, x_2, x_3 and x_4 be the number of houses built under project 1, project 2, project 3 and projects 4, respectively. The according to the conditions of the problem:

$$1 \times x_1 + 2 \times x_2 + 1 \times x_3 + 1.5 \times x_4 = 68 \quad (in\,100\, gallons)$$

$$3 \times x_1 + 4 \times x_2 + 2.5 \times x_3 + 2.5 \times x_4 = 142 \quad (in\,10,00\, cu.ft)$$

$$1 \times x_1 + 2 \times x_2 + 1.5 \times x_3 + 1 \times x_4 = 64 \quad (in\, millions)$$

$$10 \times x_1 + 10 \times x_2 + 9 \times x_3 + 8 \times x_4 = 448 \quad (in\,1000\ hours)$$

The augmented matrix is

$$A_b = \left[\begin{array}{cccc|c} 1 & 2 & 1 & 1.5 & 68 \\ 3 & 4 & 2.5 & 2.5 & 142 \\ 1 & 2 & 1.5 & 1 & 64 \\ 10 & 10 & 9 & 8 & 448 \end{array}\right]$$

$$\underset{R}{\tilde{}} \left[\begin{array}{cccc|c} 1 & 2 & 1 & 1.5 & 68 \\ 0 & -2 & -0.5 & -2 & -62 \\ 0 & 2 & 0.5 & -0.5 & -4 \\ 0 & -10 & -1 & -7 & -232 \end{array}\right] \quad by\ R_2 - 3R_1, R_3 - R_1\ and\ R_4 - 10R_1$$

$$\tilde{R} = \left[\begin{array}{cccc|c} 1 & 0 & 0.5 & -0.5 & 6 \\ 0 & -2 & -0.5 & -2 & -62 \\ 0 & 2 & 1 & -1 & -8 \\ 0 & 0 & 1.5 & 3 & 78 \end{array}\right] \quad by\, R_1 + R_2; 2R_3; R_4 + (-5)R_2$$

$$\tilde{R} = \left[\begin{array}{cccc|c} 1 & 0 & 0.5 & -0.5 & 6 \\ 0 & 1 & 0.25 & 1 & 31 \\ 0 & 0 & 1 & -1 & -8 \\ 0 & 0 & 1.5 & 3 & 78 \end{array}\right] \quad by -\frac{1}{2}R_2$$

$$\tilde{R} = \left[\begin{array}{cccc|c} 1 & 0 & 0 & 0 & 10 \\ 0 & 1 & 0 & 0 & 8 \\ 0 & 0 & 1 & 0 & 12 \\ 0 & 0 & 0 & 1 & 20 \end{array}\right] \quad by\, R_2 + (-1.25)R_4 \; and \; R_3 + R_4$$

Hence $x_1 = 10,\ x_2 = 8, x_3 = 12\, and\ x_4 = 20$ are the number of houses built for each project

CONSISTENCY CRITERION

A general system A*x* =b of *m* linear equations in *n* unknowns x_1, x_2, ...x_n or may not have a solution. Ig the system has no solution, that is, if there does not exist a vector which satisfies the equation A_x =b then the system is termed as inconsistent; In the sample case when *m=n,* that is the number of equations equals the number of unknowns and the matrix *A* of coefficient is nonsingular, the following theorem shown that the system (1) is consistent.

Theorem. The system $Ax = b,$ with $m = n$ and matrix A nonsingular, has a unique solution $x = A^{-1}b.$

Proof. Since A is nonsingular, its inverse A^{-1} exists. Multiplying both the sides of the equation, $Ax = b\, by\, A^{-1},$ we have $\left(A^1 - A\right)x = A^1 b.$

Since $A^{-1}A = l_{n,}\ the\, above$ equation becomes $l_n x = x = A^{-1}b,$ giving the solution vector $x = A^{-1}b$. To show the uniqueness, let be a second solution. Then $At = b\, and$ so $y = A^{-1}b = x$

Hence the theorem.

When the system of equations under consideration is homogeneous, that is, when we have the following:

Corollary. A **system** $Ax = 0$ of homogenous linear equations in unknowns has a unique solution $x = 0$ if and only if A is a nonsingular matrix. $x = 0$ is a trivial solution.

In case when the number of equation is not equal to the number of unknowns the system is either not consistent (so that there is no solution at all) or there may be an infinite number of solution of the system. As to when a nonhomogeneous system of equations is consistent and when it is not, the following criterion provides the answer.

Consistency Criterion. Let $Ax = b$e a system of m linear equation in n unknowns. The equations have a solution if and only if the rank of A is equal to the rank of the augmented matrix A_b.

Proof.

Since the condition, Rank $A =$ Rank A_b is always satisfied in case of a system of homogenous linear equation s, such equations therefore, necessary have a solution. We have already seen in that if Rank $A = n, and\ m = n,$ then $x = 0$ is the unique solution of $Ax = 0$.

Theorem. A system of homogenous equation in n unknows

$$a_{11}x_1 + a_{12}x_2 + \ ... + a_{1n}x_n = 0$$

$$a_{21}x_1 + a_{22}x_2 + \ ... + a_{2n}x_n = 0$$

$$\vdots \qquad \vdots \qquad ... \qquad \vdots \qquad \vdots$$

$$a_{m1}x_1 + a_{m2}x_2 + \ ... + a_{mn}x_n = 0$$ has a nontrivial solution if $m < n$.

It can be shown that, in a system of linear equation in unknowns, if Rank $A = Rank\ A_b = K < m,$ then any x which satisfies k of the equations for which the corresponding rows of in echelon from are nonzero satisfies all the equations of the set. Moreover, if $k < n, then\, n - k$ of the variable can be assigned arbitrary values and the remaining variable can be found provided that th columns od A in echelon form associated with the variable are nonzero. Naturally, therefore, in this case the solution set is infinite. We shall see in that the solution set in this case is vector space called the solution space of the system.

For a homogenous system of linear equations, we have the following necessary and sufficient condition for a nontrivial solution.

Theorem. A system of homogenous linear equations $Ax = 0$ in variable solution and only nontrivial solution if and only if the rank of A is less than

Hence a system $Ax = 0\ of\ n\, equations\, in\, n$ variables has a nontrivial solution if and only if A is singular.

Example. Examine the following homogeneous system for nontrivial solution:

$$x_1 - x_2 + 2x_3 + x_4 = 0$$

$$3x_1 + 2x_2 + \quad x_4 = 0$$

$$4x_1 + x_2 + 2x_3 + 2x_4 = 0$$

Solution. The matrix of coefficients is

$$A = \begin{bmatrix} 1 & -1 & 2 & 1 \\ 3 & 2 & 0 & 1 \\ 4 & 1 & 2 & 2 \end{bmatrix}$$

We boring into the echelon form by indicated elementary row operation.

$$A\tilde{R}\begin{bmatrix} 1 & -1 & 2 & 1 \\ 0 & 5 & -6 & -2 \\ 0 & 5 & -6 & -2 \end{bmatrix} by\, R_2 - 3R_1\ and\ R_3 - 4R_1$$

$$\tilde{R}\begin{bmatrix} 1 & -1 & 2 & 1 \\ 0 & 5 & -6 & -2 \\ 0 & 0 & 0 & 0 \end{bmatrix} by\, R_3 - R_2$$

$$\tilde{R}\begin{bmatrix} 1 & -1 & 2 & 1 \\ 0 & 1 & -\frac{6}{5} & -\frac{2}{5} \\ 0 & 0 & 0 & 0 \end{bmatrix} by\, \frac{1}{5}R_2$$

$$\tilde{R}\begin{bmatrix} 1 & 0 & \frac{4}{5} & \frac{3}{5} \\ 0 & 1 & -\frac{6}{5} & -\frac{2}{5} \\ 0 & 0 & 0 & 0 \end{bmatrix} by\, R_1 + R_2$$

The rank of the matrix is $2 < 4$. The equation has a nontrivial solution. The first two rows of the above matrix gives the following relations:

$$x_1 + \frac{4}{5}x_3 + \frac{3}{5}x_4 = 0$$

$$x_2 - \frac{6}{5}x_3 - \frac{2}{5}x_4 = 0$$

$$x_1 = -\frac{4}{5}x_3 = \frac{3}{5}x_4$$

$$\Rightarrow \begin{bmatrix} x_1 \\ x_2 \\ x_3 \\ x_4 \end{bmatrix} = \begin{bmatrix} -\frac{4}{5}x_3 - \frac{3}{5}x_4 \\ \frac{6}{5}x_3 + \frac{2}{5}x_4 \\ x_3 \\ x_4 \end{bmatrix}$$

Assigning arbitrary values to x_3, x_4, *we find* the corresponding values of x_1, x_2,. in particular, when $x_3 = a, x_4 = b$,

$$\begin{bmatrix} x_1 \\ x_2 \\ x_3 \\ x_4 \end{bmatrix} = \begin{bmatrix} -\frac{4}{5}a - \frac{3}{5}b \\ \frac{6}{5}a + \frac{2}{5}b \\ a \\ b \end{bmatrix} = a\begin{bmatrix} -\frac{4}{5} \\ \frac{6}{5} \\ 1 \\ 0 \end{bmatrix} + b\begin{bmatrix} -\frac{3}{5} \\ \frac{2}{5} \\ 0 \\ 1 \end{bmatrix}$$

Example. Three species of bacteria co-exist in a test tube and food on three foods f_1, F_2. Suppose that a bacterium of the ith species consumes, on the averages, an amount a_{ij} of the jth food per day where

$a_{11} = 1,\ a_{12} = 1,\ a_{13} = 1,\ a_{21} = 1,$

$a_{22} = 2,\ a_{23} = 3,\ a_{31} = 1,\ a_{32} = 3,\ a_{33} = 5$

Further, suppose that there are 15000, 30000, 45000 units of food of types F_1, F_2, F_3 respectively. Assuming that all food is consumed, what are the populations of the three species that can co-exist in this environment?

Solution. Let the population of the species of bacteria be $x_1, x_2 . x_3$. Since the species consumes a_{ij} units of food of types j, we have

$$\begin{bmatrix} 1 & 1 & 1 \\ 1 & 2 & 3 \\ 1 & 3 & 5 \end{bmatrix}\begin{bmatrix} x_1 \\ x_2 \\ x_3 \end{bmatrix} = \begin{bmatrix} 15000 \\ 30000 \\ 45000 \end{bmatrix}$$

That is, $x_1 + x_2 + x_3 = 15,000$

$x_1 + 2\ x_2 + 3\ x_3 = 30,000$

$x_1 + 3\ x_2 + 5\ x_3 = 45,000$

Matrix of coefficient is $A = \begin{bmatrix} 1 & 1 & 1 \\ 1 & 2 & 3 \\ 1 & 3 & 5 \end{bmatrix}$

The augmented matrix is

$$A_b = \left[\begin{array}{ccc|c} 1 & 1 & 1 & 15000 \\ 1 & 2 & 3 & 30000 \\ 1 & 3 & 5 & 45000 \end{array}\right]$$

$$\tilde{R} \left[\begin{array}{ccc|c} 1 & 1 & 1 & 15000 \\ 0 & 1 & 2 & 15000 \\ 0 & 2 & 4 & 45000 \end{array}\right] \quad by\, R_2 - R_1 \ and\ R_3 - R_1$$

$$\tilde{R} \left[\begin{array}{ccc|c} 1 & 0 & -1 & 0 \\ 0 & 1 & 2 & 15000 \\ 0 & 0 & 0 & 0 \end{array}\right] \quad by\, R_1 - R_2\, and\, R_3 - 2R_3$$

Here we notice that Rank $A = Rank\ A_b$

Thus, we may choose $x_3 = a$ arbitrarily, then

$x_1 = x_3 = a, \qquad x_2 = 15,000 - 2x_3 = 15,000 - 2a$

From the first equation we see that the total population that can co-exist 15,000

If we take $x_3 = a = 3,000.\ then\, x_1 = 3,000\, and\ x_2 = 9,000$

NETWORK FLOW PROBLEMS

Many types of real world problems involve networks along which some sort of flow is observed. A network of roads and a network of irrigation system are example of such problems. There are points in the network at which a net flow either enters or leaves the system. Head works of canals and junctions of road are the points where the flow of water and volume of traffic are studies and analyzed. The fundamental principle in the analysis of these problems is that the flow is conserved – that is the total flow entering the system at a junction must be equal to the total flow leaving the junction. Mathematical model of the problems consist of linear equations which can be solved by the method already discussed.

Example. A part of a city's road network for vehicular traffic is a shown by arrows in the following diagram.

Write the equation indicating the traffic flow given in the diagram

Show that the traffic flow along AB, CD can be expressed in terms of the traffic flow along AD,

If the stretch AD or the stretch CD is closed, then shoe that the solution to the problems is unique.

Solution. Suppose that x_1, x_2 denote the number of vehicle going along $AB, CD\, and\ AD$ respectively.

Equating the incoming traffic to the outgoing traffic at each junction, we have the following mathematical model. Here we equate the incoming and outgoing vehicles.

Thus:

At $A:\ x_1 + x_3 = 360$ (360 vehicle going along $x_1\, and\, x_3$ direction)

AT $D:\ x_1 + x_3 = 390$ (As in the case above)

AT $B:\ x_1 + 250 = x_4$

At $C: \quad x_2 + 220 = x_4$

i.e, $x_1 - x_2 = -30.$ from equation, at $B\, and\, C.$

Here, $x_1 \geq 0, \quad x_2 \geq 0, \quad x_3 \geq 0.$

The augmented matrix is

$$A_b = \left[\begin{array}{ccc|c} 1 & 0 & 1 & 360 \\ 0 & 1 & 1 & 390 \\ 1 & -1 & 0 & -30 \end{array}\right] \tilde{R} \left[\begin{array}{ccc|c} 1 & 0 & 1 & 360 \\ 0 & 1 & 1 & 390 \\ 1 & -1 & -1 & -390 \end{array}\right] by\, R_3 - R_1$$

$$\tilde{R} \left[\begin{array}{ccc|c} 1 & 0 & 1 & 360 \\ 0 & 1 & 1 & 390 \\ 0 & 0 & 0 & 0 \end{array}\right] by\, R_3 + R_2$$

So $x_1 + x_3 = 360 \;\; and \;\; x_2 + x_3 = 390.$

Take $x_3 = a.\, Then\;\, x_1 = 360 - a\;\, , x_2 = 390. - a, x_3 = a \;\; where\, a\, is$ an arbitrary integer such that $0 \leq a \leq 360.$

If the stretch AD is closed, then $x_3 = 0 \;\, and\, x_1 = 360, x_2 = 390.$

Example. The flow through a network is as shown in the diagram.

The unknowns $x_1, x_2, x_3, x_4\; x_5, x_6$ indicate the number of outgoing or incoming vehicle at the given junctions.

Put Diagram

Using the fundamental principle of network flow, solve the system.

Solution. We consider the flow of traffic at various points of the network for this, let us consider the point $A.$ the variable x_1, x_2, x_4 indicate the number of incoming and outgoing vehicle shown by the arrows. One can see that there are 200 + x_1 vehicle entering at $A.$ the number of outgoing vehicle at A is $x_2 + x_4.$ by the fundamental principle of network flow, we have the following relations.

At $A: \quad x_1 + 200 = x_2 + x_4$

Similarly, equation at other points are as follows:

At $B: x_1 + x_3 = x_5 + 100$

At $C: x_2 + x_3 = x_6 + 300$

At $D: x_4 + x_6 = x_5$

We can rewrite these equations as:

$x_1 - x_2 + 0x_3 - x_4 + 0x_5 + 0x_6 = -200$

$x_1 + 0x_2 - x_3 + 0x_4 - x_5 + 0x_6 = 100$

$0x_1 + x_2 + x_3 + 0x_4 - 0x_5 - x_6 = 300$

$0x_1 + 0x_2 + 0x_3 + x_4 - x_5 + x_6 = 0$

So, the augmented matrix is

$$A_b = \left[\begin{array}{cccccc|c} 1 & -1 & 0 & -1 & 0 & 0 & -200 \\ 1 & 0 & 1 & 0 & -1 & 0 & 100 \\ 1 & 1 & 1 & 0 & 0 & -1 & 300 \\ 0 & 0 & 0 & 1 & -1 & 1 & 0 \end{array}\right]$$

The row operations on A_b reduce the matrix to echelon form as follow:

$$A_b\tilde{R} = \left[\begin{array}{cccccc|c} 1 & -1 & 0 & -1 & 0 & 0 & -200 \\ 0 & 0 & 1 & 0 & -1 & 0 & 300 \\ 0 & 1 & 1 & 0 & 0 & -1 & 300 \\ 0 & 0 & 0 & 1 & -1 & 1 & 0 \end{array}\right] by\ R_2 - R_1$$

$$\tilde{R} = \left[\begin{array}{cccccc|c} 1 & -1 & 0 & -1 & 0 & 0 & -200 \\ 0 & 1 & 1 & 1 & -1 & 0 & 300 \\ 0 & 1 & 1 & 0 & 1 & -1 & 0 \\ 0 & 0 & 0 & 1 & -1 & 1 & 0 \end{array}\right] by\ R_3 - R_2$$

$$\tilde{R} = \left[\begin{array}{cccccc|c} 1 & 0 & 1 & 0 & -1 & 0 & 100 \\ 0 & 1 & 1 & 1 & -1 & 0 & 300 \\ 0 & 0 & 0 & -1 & 1 & -1 & 0 \\ 0 & 0 & 0 & 0 & 0 & 0 & 0 \end{array}\right] by\ R_1 + R_2\ and R_4 + R_3$$

So, the network flow system becomes

$x_1 + x_2 \quad - x_5 = 100$

$x_2 \; + x_2 \qquad\qquad - x_6 = 300$

$x_4 - x_5 + x_6 = 0$

We can give arbitrary values to $x_3, x_5, x_6 as:$

$x_3 = r, \quad x_5 = s, \quad x_6 = t$

Then, from, we have

$x_4 = x_5 - x_6 \;= s - t$

So, becomes

$x_1 = 100 - x_3 + x_5$

$= 100 - r + s$

$x_2 = 300 - r + t$

$x_3 = r$

$x_4 = s - t$

$x_5 = s$

$x_6 = t$

The values of $x_i, i = 1, 2, \ldots, 6$ given above provide the most general solution of network system.

Obviously, each arbitrary value of r, s and t gives a solution of the system

In particular if $x_3 = 100 \;,\; x_5 = 50 \;,\; x_6 = 50, then$

$x_1 = 100 - x_3 + x_5 = 100 - 100 + 50 = 50$

$x_2 = 300 - x_3 + x_6 = 300 - 100 + 50 = 250$

$x_4 = x_5 - x_6 = 50 - 50 = 0$

The particular solution is

$x_1 = 50$

$x_2 = 250$

$x_3 = 100$

$x_4 = 0$

$x_5 = 50$

$x_6 = 50$.

Proof of Theorem. The matrix of the coefficient and the augmented matrix are

$$A = \begin{bmatrix} a_{11} & a_{12} \cdots & a_{1n} \\ a_{21} & a_{22} \cdots & a_{2n} \\ \vdots & \vdots \quad \cdots & \vdots \\ a_{m1} & a_{m2} \cdots & a_{mn} \end{bmatrix}, A_b = \left[\begin{array}{cccc|c} a_{11} & a_{12} \cdots & & a_{1n} & b_1 \\ a_{21} & a_{22} & \cdots & a_{2n} & b_2 \\ \vdots & \cdots & \vdots & \cdots & \vdots \\ a_{m1} & am_2 & \cdots & a_{mn} & b_m \end{array}\right]$$

We shall represent the columns of the matrices by bold faced letters; thus a_j *will be the* jth columns of the matrix. A Now the ranks of A and A_b *are* the maximum number of linearly independent columns in A and A_b, represent. Since every column of A is also a column of A_b the rank of A cannot exceed the rank of A_b. There are, therefore, only two possibilities:

(i) Rank $A < Rank\ A_b$

(ii) Rank A = Rank A_b

We note that the rank of A_b cannot be greater than $1 + Rank\ A.$ if the first possibility is true, then any set of linearly independent columns of A_b must include the column b so that b is not a linear combination of the columns of A. For otherwise , $Rank\ A = Rank\ A_b$. This it follows that there do not exist any x_j's satisfying

$$\sum_{j=1}^{n} x_j a_j = b,$$

Where a_j represent the jth column of $A,$ that is, there exist no $x_j^{'}s$ satisfying $Ax = b.$ Hence there is no solution and given system is inconsistent.

Next consider the second possibility that $Rank\ A = Ran\ A_b = k.$ in this case there is a set of k linearly independent columns of $A.$ without any loss of generality, we can suppose that the first k columns of A are linearly independent. Then every other column of A_b is expressible as a linear combination of these k columns. Thus there exist numbers $x_1, x_2, \ldots, x_k$ such that

$$\sum_{j=1}^{n} x_j a_j = \mathbf{b},$$

Adding zero times the remaining $n-k$ columns of $A,$ we get $Ax = b$, where $x = (x_1, x_2, \ldots, x_k, 0, \ldots, 0).$ Hence in this case there is at least one solution.

Conversely, if a solution $y = (y_1, y_2, \ldots, y_n)$ of the system of equations $Ax = b$ exist then b must be a linear combination of columns of $A.$ Thus there exist scalar $y_j{'}$ not all zero, such that

$$\sum_{j=1}^{n} y_j a_j = \mathbf{b}, \quad or\ Ay = \mathbf{b}.$$

Taking transpose of both sides, we have $y^T A^T = b^T.$

Thus $\mathbf{b}^T$ is a linear combination of the rows of A^T . Hence

$Rank\ A_b = Rank A_b^T = Rank A^T = Rank\ A.$

The theorem is proved.

Proof of Theorem

The proof of the theorem will be given by induction on

If $, = m = 1,$ there is only one equation

$a_{11}x_1 + a_{12}x_2 + \ldots + a_{1n}x_n = 0$

With $n \geq 2$. if $a_{11} = 0,$ a nontrivial solution of (1) is

$x_1 = 1,\ x_2 = x_3 = x_3 = \ldots = x_n = 0.$

If $a_{11} \neq 0,$ we take $x_2 = \ldots = x_n = 1$ So that (1) becomes

$a_{11}x_1 + a_{12} + \ldots + a_{1n} = 0$

Giving $$x_1 = -\frac{1}{a_{11}}\left(a_{12} + \ldots + a_{1n}\right), x_2 = \ldots = x_n = 1,$$

Which is a nontrivial solution of (1).Hence the theorem is true for $m = 1\ and\, n \geq 2.$

Now assume that the theorem is true for some positive integer $m = k.We$ shall show that it is true for $m = k + 1 and\, k + 1 < n.$

Consider the set of equations

$$\left.\begin{array}{l} a_{11}\ x_1 + a_{12}\,x_2 + \cdots + a_{1n}\,x_n = 0 \\ a_{21}\ x_1 + a_{22}\ x_2 + \ldots + a_{2n}\ x_n = 0 \\ \vdots \qquad \vdots \qquad \ldots \qquad \vdots \\ a_{k1}\ x_1 + a_{k2}\ x_2 + \cdots + a_{kn}\ x_n = 0 \\ a_{k+11}\,x_1 + a_{k+12}\,x_2 + \ldots + a_{k+1n}\,x_n = 0 \end{array}\right\}$$

If $a_{11} = a_{21} = \ldots = a_{k1} = a_{k+11} = 0,$ then

$x_1 = 1,\ x_2 = 0 = x_3 = \ldots = x_n$

is a nontrivial solution of the system. Hence, we may assume that at least of the coefficient is nonzero. Without any loss of generality, let $a_{11} \neq 0,$ for otherwise we can consider the equation with nonzero coefficient of and reliable it as the first equation. Then

$$x_1 = -\frac{1}{a_{11}}\left(a_{12}x_2 + a_{13}x_3 + \ldots + a_{1n}x_n\right).$$

Substituting this value of x_1 in the second through the $(k+1)$st equations of the system, we obtain the equivalent system.

$$\left.\begin{array}{l} a_{11}\ x_1 + a_{12}\,x_2 + a_{13}\ x_3 + \cdots + a_{1n}\,x_n = 0 \\ \quad b_{22}\ x_2 + b_{23}\ x_3 + \ldots + b_{2n}\ x_n = 0 \\ \quad \vdots \qquad \vdots \qquad \ldots \qquad \vdots \\ b_{k2}\ x_2 + b_{k3}\ x_3 + \cdots + b_{kn}\ x_n = 0 \\ a_{k+12}\,x_2 + b_{k+13}\,x_3 + \ldots + b_{k+1n}\,x_n = 0 \end{array}\right\}$$

Where

$$b_{ij} = a_{ij} - \frac{a_{i1}a_{ij}}{a_{11}}, i = 2,3,\ldots,k+1, j = 2,3,\ldots,n.$$

The last k equation of the system are in $(n-1)$ unknowns $x_2, x_3, \ldots, x_n$ and $k < n-1$, since $k+1 < n$. Therefore, by the induction hypothesis, these equations have a nontrivial solution $x_2, x_3, \ldots, x_n$ $Once\, x_2, x_3, \ldots, x_n$ are known, we obtain

$$x_1 = -\frac{1}{a_{11}}\left(a_{12}x_2 + a_{13}x_3 + \ldots + a_{1n}x_n\right).$$

Thus, the theorem is true for $m = k+1 < n$ and hence for all $m < n$, by induction.

Proof of Theorem. Let the rank of A be k such that $k < n$. There are K linearly independent columns, and every set of $k+1$ columns of A is linearly dependent if we transfer these columns to be the first k columns, then every remaining column in linearly dependent on these so that there exit scalars $x_1, x_2, \ldots, x_k$, not all zero such that $\sum_{j=1}^{n} x_j a_j = a_{k+1}$ and $i = 1,2,\ldots,n-k$.

Thus $(x_1, x_2, \ldots, x_k, 0, 0, \ldots, 0)$ is a nontrivial solution of $A_x = 0$. In fact the last $n-k$ element of the solution vector can be given arbitrary values.

Conversely, if there is a nontrivial solution $y \neq 0$ of $AX = 0\, and\ Rank\ A = n$, then A^{-1} exist and so, from $Ay = 0$, we obtain $A^{-1}\,Ay = A^{-1}\,0 = 0$

i.e., $y = 0$ which is contradiction. Hence Rank A is less than n.

SUMMARY AND DISCUSSION

In this chapter, we discussed about system of linear equation and different methods to solve system of linear equation.

System of Linear Equations. An equation of the form $a_1x_1 + a_2x_2 + a_3x_3 + \ldots + a_nx_n = b$ is called a linear equation in the variable $x_1, x_2, x_3, \ldots, x_n$. Here $a_1, a_2 \ldots, a_n$ b are constants, usually real numbers A collection of several such equation is called a system of linear equations. The theory of matrices has been usefully employed in various branches of well as applied mathematics. In what follow, we shall apply this theory to the solution of m linear equation in n unknowns.

Let $Ax = b$ where A ,x ,b are the matrices as given below:

$$A = \begin{bmatrix} a_{11} & a_{12} \ldots & a_{in} \\ a_{21} & a_{22} \ldots & a_{2n} \\ \vdots & \vdots & \ldots \vdots \\ a_{m1} & a_{m2} \ldots & a_{mn} \end{bmatrix}, \quad x = \begin{bmatrix} x_1 \\ x_2 \\ \vdots \\ x_n \end{bmatrix}, \quad b = \begin{bmatrix} b_1 \\ b_2 \\ \vdots \\ b_m \end{bmatrix}$$

The matrix A is called matrix of coefficients of the system of equations (1), the column of constants b_i forms column vector b and the unknowns x_i from the column vector x.

Nonhomogeneous Linear Equations. The equation of the type given in (1) above with $b \neq 0,$ are called system of nonhomogeneous linear equations.

Homogeneous Linear Equations. If $b = 0$, then the system of equations (1) is known as a homogenous linear system.

Solution of the system. An order $n - tuple\, y = (y_1, y_2, \ldots, y_n)$ is called a **solution** of the system (1), if, whenever $x_1, x_2 \ldots, x_{nn}$ are replaced by $y_1, y_2, \ldots y_n$ respectively in each equation of (1, the resulting statements are true statements. The vector y is then also said to satisfy the system of equations.

Augmented Matrix. The matrix

$$[A|B] = A_b = \left[\begin{array}{cccc|c} a_{11} & a_{12} & \cdots & a_{1n} & b_1 \\ a_{21} & a_{22} & \cdots & a_{2n} & b_2 \\ \vdots & \vdots & \cdots & \vdots & \vdots \\ a_{m1} & a_{m2} & \cdots & a_{mn} & b_m \end{array}\right]$$

is known as the augmented matrix of the system of linear equations. It is quite important in the sense that its rank and the rank of A (the matrix of coefficients) determine whether the system of equations $\mathbf{Ax} = \mathbf{b}$ does or does not have a solution.

METHODS TO SOLVE SYSTEM OF LINEAR EQUATIONS

Gaussian elimination method

One of the several methods employed to solve a system of n linear equations in n variable, is known as Gaussian Elimination Procedure (named after its discoverer, the German mathematician C.F. Gauss, 1777 -1855).

Gauss-Jordan Elimination Method

While reducing the given system of linear equation in to the equivalent system (II), we eliminated $x_k, k = 1,2 ,\ldots,n-k$ from the $(K+1)th,(K+2)th\ldots nth$ equations only, if we remove x_k from the first $(k-1)$ equations also then x_k will appear only in kth equation. With this modification we get another equivalent system from the given system expressing each x_k explicitly and we no more require any backward substitution of these for finding the values of other unknowns, this process is called the Gauss-Jordan Elimination Method.

CHAPTER 3

DETERMINANTS

INTRODUCTION

The reader may be familiar with the concept of determinants. This concept played a basic role in the solution of systems of linear equations before the use of matrices and computers. The use of determinants is becoming less and less with every passing day. Nevertheless, determinants still play an important but minor in finding solution of linear problems.

This chapter include concept of determinants still of a square matrix, minors' role in finding solution of linear problems.

This chapter includes concept of determinant of a square matrix, minors and cofactors, properties of determinants, Laplace expansion of a determinant, ad joint and inverse of a matrix.

Determinant of a square Matrix

Definition. Method for solving systems of homogeneous and nonhomogeneous linear equations were given. We shall make use of those methods to give formal definition of a determinant. Before taking up the general case, we consider the following special case of two linear equations:

$$a_{11}x_1 + a_{12}x_2 = b_1$$

$$a_{21}x_1 + a_{22}x_2 = b_2$$

In two unknowns x_1, x_2. Here

$$A = \begin{bmatrix} a_{11} & a_{12} \\ a_{21} & a_{22} \end{bmatrix}, \quad A_b = \left[\begin{array}{cc|c} a_{11} & a_{12} & b_1 \\ a_{21} & a_{22} & b_2 \end{array}\right]$$

are the coefficient matrix and the augmented matrix respectively with entries as real as real numbers. By the Gauss-Jordan procedure, the solution of these equations is

$$x_1 = \frac{(b_1 a_{22} - b_2 a_{12})}{(a_{11}a_{22} - a_{21}a_{12})}, x_2 = \frac{(b_1 a_{11} - b_1 a_{21})}{(a_{11}a_{22} - a_{21}a_{12})}$$

Provided that

$$a_{11}a_{22} - a_{21}a_{12} \neq 0$$

The scalar $a_{11}a_{22} - a_{21}a_{12}$ is uniquely determined by the matrix A. It is called the determinants of order 2 of the square matrix A given in and is denoted by det A or $|A|$. Thus.

$$|A| = \det A = a_{11}a_{22} - a_{21}a_{12} = \begin{vmatrix} a_{11} & a_{12} \\ a_{21} & a_{22} \end{vmatrix}$$

Note the two vertical bars instead of square brackets use of matrices.

We note that a matrix A of order 2. So, in this case, we may regard as a function whose domain is the set of all square matrix of order 2 and whose range is a subset of real numbers.

The following properties of determinants of order 2 follow directly from the above definition.

(i) For any real number

$$\begin{vmatrix} k\, a_{11} & a_{12} \\ k\, a_{21} & a_{22} \end{vmatrix} = \text{k} \begin{vmatrix} a_{11} & a_{12} \\ a_{21} & a_{22} \end{vmatrix}$$

(ii) If $a_2 = b_{12} + c_{12}.\ a_{22} = b_{22} + c_{22}, then$

$$\begin{vmatrix} a_{11} & a_{12} \\ a_{21} & a_{22} \end{vmatrix} = \begin{vmatrix} a_{11} & b_{12} + c_{12} \\ a_{21} & b_{22} + c_{22} \end{vmatrix}$$

$$= \begin{vmatrix} a_{11} & a_{12} \\ a_{21} & a_{22} \end{vmatrix} + \begin{vmatrix} a_{11} & c_{12} \\ a_{21} & a_{22} \end{vmatrix}$$

(iii) If the two columns of idenitical the $\det A = 0.That$ is

$$\begin{vmatrix} a_{11} & a_{12} \\ a_{21} & a_{22} \end{vmatrix} = 0$$

(iv). The determinant of the unit matrix is 1. That is

$$\begin{vmatrix} 1 & 0 \\ 0 & 1 \end{vmatrix} = 1$$

It will be seen later that these properties are characteristics of a determinants of square matrix of any order.

Definition. (Determinants of Order). Firstly we define the determinant of an $n \times n$ matrix inductively, That is, from our knowledge of a determinant of order 2, we define a determinant of order 3 and use this definition of a determinant of order 3 to describe a determinant of order 4 and so on.

Thus a determinant of order 3 is defined as follows:

Let

$$A = \begin{bmatrix} a_{11} & a_{12} & a_{13} \\ a_{21} & a_{22} & a_{23} \\ a_{31} & a_{32} & a_{33} \end{bmatrix}$$

Then

$$\det A = |A| = a_{11} \begin{vmatrix} a_{22} & a_{23} \\ a_{32} & a_{33} \end{vmatrix} - a_{12} \begin{vmatrix} a_{21} & a_{23} \\ a_{31} & a_{33} \end{vmatrix} + a_{13} \begin{vmatrix} a_{21} & a_{23} \\ a_{31} & a_{32} \end{vmatrix}$$

Note the minus sign before the second term on the right-hand side of (1)

For example, if

$$A \begin{bmatrix} 4 & 1 & 2 \\ 3 & 2 & 5 \\ 1 & 2 & 3 \end{bmatrix}, then$$

$$\det A = \begin{vmatrix} 4 & 1 & 2 \\ 3 & 2 & 5 \\ 1 & 2 & 3 \end{vmatrix} = 4\begin{vmatrix} 2 & 5 \\ 2 & 3 \end{vmatrix} - 1\begin{vmatrix} 3 & 5 \\ 1 & 3 \end{vmatrix} + 2\begin{vmatrix} 3 & 2 \\ 1 & 2 \end{vmatrix}$$

$$= 4(6-10) - 1(9-5) + 2(6-2)$$

$$= -16 - 4 + 8 = -12$$

There is a simple method to calculate a determinant of order 3.

$$\det A = a_{11}(1_{22}a_{33} - a_{23}a_{32}) - a_{12}(a_{21}a_{33} - a_{23}a_{31}) + a_{13}(a_{21}a_{32} - a_{22}a_{31})$$

$$= a_{11}a_{22}a_{33} - a_{12}a_{23}a_{31} + a_{13}a_{21}a_{32} - a_{11}a_{23}a_{32} - a_{12}a_{21}a_{33} - a_{13}a_{22}a_{31}$$

We write the columns of and adjoin to these the first two columns as below:

Now calculate the six products of number on the directed lines taking plus sign with those products on arrows pointing downwards and minus sign with products on arrows pointing upwards. Adding these products we get the value of $\det A.$ This is known as Sarrus's rule

Note: The method of arrows given above works only for $n = 2, 3,$ and not for $n \geq 4.$

Definition. Let be a square matrix of order $n.$ The matrix obtained from A by deleting its row and column is again a matrix M_{ij} of order $n-1. M_{ij}$ is called the *ij th* minor of $A.$[1]

Definition. Let M_{ij} be the mi nor of a square matrix A of order $n.$ Then

$$A_{ij} = (-1)^{i+j} \det M_{ij}$$

Is called the $ijth$ cofactor of $A.$

Observe that the sign on the right hand side of the above equality is positive or negative according as $i + j$ is even or odd.

Example 1. Let

$$A = \begin{bmatrix} 2 & -1 & 3 & 4 \\ 3 & 2 & 5 & -2 \\ 3 & 4 & 2 & 1 \\ -3 & 2 & 5 & 1 \end{bmatrix}$$

Then

$$M_{23} = \begin{bmatrix} 2 & -1 & 4 \\ 3 & 4 & 1 \\ -3 & 2 & 1 \end{bmatrix}, \quad M_{42} = \begin{bmatrix} 2 & 3 & 4 \\ 3 & 5 & -2 \\ 3 & 2 & 1 \end{bmatrix}$$

1. Some author call M_{ij} of the matrix M_{ij} as the mirror of the $(i, j)th$ element.

and $A_{23} = (-1)^{2+3} \det M_{23} = -82$

$A_{42} = (-1)^{4+2} \det M_{42} = -45$

Consider the expansion of given in. Using the fact

$$A_{11} = (-1)^{1+1} \begin{vmatrix} a_{22} & a_{23} \\ a_{32} & a_{33} \end{vmatrix}, \quad A_{12} = (-1)^{1+2} \begin{vmatrix} a_{21} & a_{23} \\ a_{31} & a_{33} \end{vmatrix}$$

$$A_{13} = (-1)^{1+3} \begin{vmatrix} a_{21} & a_{22} \\ a_{31} & a_{32} \end{vmatrix}$$

We may write (1) as

$\det A = a_{11} A_{11} + a_{12} A_{12} + a_{12} A_{12} + a_{13} A_{13}.$

We now define det of a matrix A of order n as follows:

Definition. For the matrix $A = \begin{bmatrix} a_{11} & a_{12} \dots a_{1n} \\ a_{21} & a_{22} \dots a_{2n} \\ \dots & \vdots \dots \vdots \\ a_{n1} & a_{n2} \dots a_{nn} \end{bmatrix}$ of order det A by

$\det A = a_{i1} A_{i1} + a_{i2} A_{i2} + \ldots + a_{in} A_{in}.$

The expression on the right-hand side of (1) is an expansion of by cofactor of the row of This is called Laplace's Expansion of a determinant of order The definition (1) given above indicates that the determinant of a matrix can be evaluated by any row of the matrix. We only have to be careful about the sign of the cofactor of the corresponding element.

Example. Let

$$A = \begin{bmatrix} 3 & 2 & 1 & -1 \\ 4 & 5 & 1 & 2 \\ -2 & 3 & 0 & 1 \\ 2 & 1 & 3 & 5 \end{bmatrix}$$ Then

$$\det A \begin{vmatrix} 3 & 2 & 1 & -1 \\ 4 & 5 & 1 & 2 \\ -2 & 3 & 0 & 1 \end{vmatrix}$$

$$= 3\begin{vmatrix} 5 & 1 & 2 \\ 3 & 0 & 1 \\ 1 & 3 & 5 \end{vmatrix} - 2\begin{vmatrix} 4 & 1 & 2 \\ -2 & 0 & 1 \\ 2 & 3 & 5 \end{vmatrix} + 1\begin{vmatrix} 4 & 5 & 2 \\ -2 & 3 & 1 \\ 2 & 1 & 5 \end{vmatrix}$$

$$+1\begin{vmatrix} 4 & 5 & 2 \\ -2 & 3 & 0 \\ 2 & 1 & 3 \end{vmatrix}$$

$$= 2\left[5(-3) - 1(14) + 2(9)\right] - 2\left[4(-3) - 1(-12) + 2(-6)\right]$$

$$1\left[4(14) - 5(-12) + 2(-8)\right] + 1\left[4(9) - 5(-6) + 1(-8)\right]$$

$$= 3(-15 - 14 + 18) - 2(-12 + 12 - 12) + 1(56 + 60 - 16) + 1(36 + 30 - 8)$$

$$= -33 + 24 + 100 + 58 = 149$$

Remark. Although the above technique to evaluate determinant of an matrix seems quite straightforward, yet, in practice, it is very laborious to work with when $n > 3$ because it involves a lot of calculations. We shall later discus methods which simplify these calculations.

Note that, for some special type of matrices, their determinants can be easily evaluated. For instance, if a matrix is triangular (upper or lower), that is, all its entries below or above the main diagonal are zero respectively, then its determinant is just the product of the elements on the main diagonal.

Example. Let

$$A = \begin{bmatrix} a_{11} & 0 & 0 & 0 \\ a_{21} & a_{22} & 0 & 0 \\ a_{31} & a_{32} & a_{33} & 0 \\ a_{41} & a_{42} & a_{43} & a_{44} \end{bmatrix}$$

Then $\det A = a_{11}A_{11} + a_{12}A_{12} + a_{13}A_{13} + a_{14}A_{14}$

$$= a_{11}A_{11} + 0A_{12} + 0A_{13} + 0A_{14} = a_{11}A_{11}$$

$$= a_{11}\begin{vmatrix} a_{22} & 0 & 0 \\ a_{32} & a_{33} & 0 \\ a_{42} & a_{43} & a_{44} \end{vmatrix} = a_{11}\, a_{22}\begin{vmatrix} a_{33} & 0 \\ a_{43} & a_{44} \end{vmatrix} = a_{11}a_{22}a_{23}a_{44}$$

By induction on it is easy to generalize the above result to a triangular matrix of any order.

Similarly, for an upper triangular matrix. Thus

Theorem. Let $A = \left[a_{ij}\right]$ be an $n \times n$ triangular matrix. Then $\det A$ is the product of the elements on the main diagonal, that is,

$\det A = a_{11}a_{22}\, a_{33} \ldots a_{nn}$

Note: In general, to evaluate the determinant of a matrix one should start with the row (or column) of A with the maximum number of zero entries.

AXIOMATIX DEFINITION OF A DETERMINANT

Notation. Let M_n be the set of all nxn matrices with entries from F, where F is the field of real or complex numbers, An element A of M_n is an $n \times n$ matrix

$$A = \begin{bmatrix} a_{11} & a_{12} & \cdots & a_{1n} \\ a_{21} & a_{22} & \cdots & a_{2n} \\ \vdots & \vdots & \cdots & \vdots \\ a_{n1} & a_{n2} & \cdots & a_{nn} \end{bmatrix}$$

If we denote the jth column

$a1j$

a_{2j}

$\vdots$

a_{nj}

Of ba $\boldsymbol{a}_j$ then we can write as

$$A = \begin{bmatrix} \boldsymbol{a}_1 & \boldsymbol{a}_2 & \cdots & \boldsymbol{a}_j & \cdots & \boldsymbol{a}_n \end{bmatrix}$$

The matrix

$$A^* = \begin{bmatrix} a_1 & a_2 & \cdots & ca_j \cdots & a_n \end{bmatrix}$$

Denotes the matrix obtained from A by multiplying every element of its column by a scalar

Similarly, the matrix

$$A^{**} = \begin{bmatrix} a_{11} & a_{12} & \dots & a_{1i} + ca_{1j} & \dots & a_{1j} & \dots & a_{1n} \\ a_{21} & a_{22} & \dots & a_{2i} + ca_{2i} & \dots & a_{2j} & \dots & a_{2n} \\ \vdots & \vdots & \dots & \vdots & \dots & \vdots & \dots & \vdots \\ a_{n1} & a_{n2} & \dots & a_{ni} + ca_{nj} & \dots & a_{nj} & \dots & a_{nn} \end{bmatrix}$$

We now give an axiomatic definition of determinant of a square matrix of order n as follows:

Definition. A function $\det M_n \to F\, which$ associates with a matrix A in M_n a unique element det $A\, of\, F,$ is said to be the determinant function if it satisfies the following conditions:

(i) det is a linear function of column of $A.$ That is, if

$$a_j = \beta b_j + \gamma c_j, for\, \beta, \gamma \in F, \quad 1 \le j \le n,\, then$$

$$detA = \det\begin{bmatrix} a_1 & a_2 & \cdots & a_j & \cdots & a_n \end{bmatrix}$$

$$= \det\begin{bmatrix} a_1 & a_2 \cdots & \beta b_j & +\gamma c_j & \cdots a_n \end{bmatrix}$$

$$= \beta \det\left[a_1 \quad a_2 \ldots b_j \ldots a_n\right] + \gamma \det\left[a_1 \quad a_2 \ldots c_j \ldots a_n\right].$$

(ii) If any two adjacent columns of A are identical then $det\, A = 0$

(ii) $det\, I_n = 1. Here\, I_n\, is\, the\, n \times n$ identity matrix and 1 is the multiplicative identity of F.

Definition. The element $\det A\, of\, F$ associated with a matrix $A. \in M_n$ is called the determinants of A.

Note that while a, matrix A is a symbol, its determinant $\det A\; is\, a$ real or complex number according as the entries of $A. are\; from\; \boldsymbol{R}\; or\; \boldsymbol{C}. a_i\;\; by\, b_i A$

Henceforth, for the sake of simplicity's and convenience, we shall make no distinction between the properties of the determinant function and those of determinants.

In the following theorem, we give a few properties of the determinant function.

Theorem. Let A be a square matrix of order n. then, the determinant function has the following properties.

(i) If the column vector $\boldsymbol{a}_i$ of A is of the form $a_i = a\,\boldsymbol{b}_i$. Then $\det A = \alpha \det B, (\alpha\, being\;\; a\, scalar)$, Where B is the matrix obtained from by replacing its ith column.

(ii) If a column $a_i\;\, of\;\, A\, is\, 0,$ then $\det A = 0$

(iii) If a column $a_i\, of\;\, A$ is of the form $a_i = b_i + c_i$ then $\det A = \det B + \det C$

where B and C are matrices obtain from A by replacing its column by b_i and c_i respectively .

(iv) if a_i is replaced by $a_i + c\, a_{i+1}$ in A, then $\det A$ remains unchanged. That is

$$= \det\left[\boldsymbol{a}_1 \;\; \boldsymbol{a}_2 \ldots \;\; \boldsymbol{a}_1 + \boldsymbol{c}\, \boldsymbol{a}_{i+1} \;\; \boldsymbol{a}_{i+1} \ldots \boldsymbol{a}_n\right]$$

$= \det[\boldsymbol{a}_1 \ \ \boldsymbol{a}_2 \ \dots \ \ \boldsymbol{a}_1 \, \boldsymbol{a}_{i+1} \ \dots a_n]$

(v) Interchanging of any two adjacent columns of $A change$ the sign of $det \ \ A$

(vi) Interchanging of any two columns of A results in the change of sign of det A

(vii) If any two columns of A results in the change of sign of det A.

(Viii) A Constant multiplied of column of A may be added to any other column without changing the value of $\det A, That \, is$

$det = [a_1 \ \ a_2 \ \dots \ a_1 + ca_j \dots \ a_n]$, for any scalar c

$= \det[a_1 \ a_2 \ \dots \ a_i \dots \ a_n]$

Proof. (i) since by Definition, det is a linear function of A,

$\det A \ = \det[a_1 \quad a_2 \ \dots \ a_i \ \dots \ a_n]$

$\det A \ = \det[a_1 \quad a_2 \ \dots \ ab_i \ \dots \ a_n]$

$= \alpha \det[a_1 \ a_2 \dots \ b_i \dots a_n] = \alpha \det B.$ e

(ii) if each entry of a column $a_{i,}$ say, is zero, then

$a_i = 0 = 0 b_i$ for any b_i

We have

$\det A = 0 \ \det B = 0$

(iii) Suppose that, for column a_i of $A, \ a_i = b_i + c_i$

Then, by taking $\beta = \gamma = 1$, we have

$\det A = \det[a_1 \ \ a_2 \dots \ b_i + c_i \ a_{i+1} \dots a_n]$

$= \det[a_1 \ \ a_2 \dots \ b_i \ a_{i+1} \dots a_n]$

$+ \det[a_1 \ \ a_2 \dots c_i \quad a_{i+1} \dots a_n]$

$= \det A + c0 = \det C,$

(v) Let $A=\left[a_1 \quad a_2 \ldots a_i\, a_{i+1} \ldots a_n\right]$

and $B=\left[a_1 \quad a_2 \ldots a_{i+1} a_i \ldots a_n\right]$

be the matrix obtained from A by interchainging its jth and $(i+1)th$ columns. Then

$$\det B = \det[a_1\ a_2\ \ldots\ a_{i+1}\ a_i\ \ldots\ a_n]$$

$$= \det[a_1\ a_2\ \ldots\ a_{i+1}\ a_i\ a_{i+1}\ \ a_{i+2}\ \ldots\ a_n]$$

$$\downarrow$$

$$= \det[a_1\ a_2\ \ldots\ a_i +\ a_i + a_{i+1}\ a_i + a_{i+1}\ \ a_{i+2}\ \ldots\ a_n]\ \text{by(iv)}$$

$$= det\ [a_1\ a_2\ \ldots\ -a_i +\ a_i + a_{i+1}\ a_i\ + a_{i+1}\ a_{1+2} \ldots a_n],\ \text{by(iv)}$$

$$= -\det[a_1\ \ a_2\ \ldots\ a_i\ a_i + \ a_{i+1}\ a_{i+2} \ldots a_n],$$

$$= \det[a_1\ a_2 \ldots a_i\ \ a_i\ + a_{i+1}\ \ldots a_n],$$

$$= -\det A$$

(vi) Suppose that two columns $a_i, a_j, i \neq j\, of A$ are identical, $, i.e.,\ a_i = a_j$ then

$$\det A = \det\left[a_1\ a_2\ \ldots\ a_i \ldots\ a_j\ \ldots a_n\right]$$

$$= -\det\left[a_1\ a_2\ \ldots a_i\ \ldots\ a_j\ \ldots a_n\ \right],\ because\ a_i = a_j$$

$$= -\det[a_1\ a_2\ \ldots\ a_i\ \ldots\ \ a_n],\ because\ \boldsymbol{a_i} = \boldsymbol{a_j}$$

$$= -\det A.$$

Hence $2 \det A = 0. That\, is, \det A = 0.$ The other properties may be verified similarly

DETERMINANT AS SUM OF PRODUCT OF ELEMENTS

Definition. Let $A=\left[a_{ij}\right]$ be a square matrix of order n with entries from a field $F.$ That the sum

$$\sum_{\sigma}(-1)^k a_{1k_1}\, a_{2k_2} \ldots a_{nk_n}$$

where $k \ is 0 or 1$ according as the permutation

$$\sigma = \begin{pmatrix} 1 & 2 & 3 & \ldots n \\ k_1 & k_2 & k_3 & \ldots k_n \end{pmatrix}$$

In S_n is even or odd, is called the determinants of A and is denoted by det $A or by$

$$\begin{vmatrix} a_{11} & a_{12} & \cdots & a_{1n} \\ a_{21} & a_{22} & \cdots & a_{2n} \\ \vdots & \vdots & \cdots & \vdots \\ a_{n1} & a_{n2} & \cdots & a_{nn} \end{vmatrix}$$

In particular, if $n = 3$, then $S_3 \ has \ 3! = 6$ elements which are

$$I = \begin{pmatrix} 1 & 2 & 3 \\ 1 & 2 & 3 \end{pmatrix}, \quad \sigma = \begin{pmatrix} 1 & 2 & 3 \\ 2 & 3 & 1 \end{pmatrix}, \sigma_2 = \begin{pmatrix} 1 & 2 & 3 \\ 3 & 1 & 2 \end{pmatrix}$$

$$\sigma_3 = \begin{pmatrix} 1 & 2 & 3 \\ 2 & 1 & 3 \end{pmatrix}, \sigma_4 = \begin{pmatrix} 1 & 2 & 3 \\ 1 & 3 & 2 \end{pmatrix}, \sigma_5 = \begin{pmatrix} 1 & 2 & 3 \\ 3 & 2 & 1 \end{pmatrix}$$

Of the above six permutations, the first three are even and the next three are odd. So if

$$A = \begin{bmatrix} a_{11} & a_{12} & a_{13} \\ a_{21} & a_{22} & a_{23} \\ a_{31} & a_{32} & a_{33} \end{bmatrix}$$

Then $\det A = \sum_{\sigma} (-1)^k \ a_{1i_1} \ a_{2i_2} a_{3i_3}$

$$= a_{11}a_{22}a_{33} + a_{12}a_{23}a_{33} + a_{13}a_{21}a_{32} - a_{11}a_{23}a_{32} - a_{12}a_{21}a_{33} - a_{13}a_{22}a_{31}$$

Here

$$\sigma = \begin{pmatrix} 1 & 2 & 3 \\ i_1 & i_2 & i_3 \end{pmatrix}$$

is one of the permutations given above, Thus, for writing the products in

(2), we first take the subject i in a_{ij} as in the first row of the corresponding permutation and j as the corresponding element below i in the second row of the permutation. The process repeated for all the six permutations For example,, the term in sum corresponding to the first permutation namely

$$I = \begin{pmatrix} 1 & 2 & 3 \\ 1 & 2 & 3 \end{pmatrix}$$

is the product $a_{11}a_{22}a_{33}$ with minus sign because σ_3 is an odd permutation.

Similarly, if $A = \begin{bmatrix} 1 & 6 & 7 \\ 2 & 3 & 5 \\ 4 & 1 & 7 \end{bmatrix}$, then, writing the corresponding values of a_{ij} in the matrix A, wc have

$$\det A = 1x3x7 + 6x5x4 + 7x2x1 - 6x2x7 - 1x5x1 - 7x3x4$$

$$= 21 + 120 + 14 - 84 - 5 - 84$$

$$= 155 - 173 = -18$$

In the following theorem we shall see that all statement about columns of matrix A in the definition of det A hold equally good for the row of A.

Theorem. Let A be the transpose of a square matrix A of order n. Then $\det A = \det A$.

Proof. Let $A = \left[a_{ij}\right] be\ an\, nx\, n\, matrix\, and$

$$A^T = \left[b_{ij}\right],\ b_{ij} = a_{ji}$$

Be the transpose of A. Here a_{ji} is the element in the ith row and jth column $of\ A^T$. Now

$$\det A^T = \sum_{\sigma} (\pm) b_{1i_1} b_{2i_2} \dots b_{ni_n}$$

$$= \sum_{\sigma} (\pm) b_{1j} b_{2j_2} \cdots b_{nj_n}$$

Where if

$$\sigma = \begin{pmatrix} 1 & 2 & 3 \ldots & n \\ j_1 & j_2 & j_3 \cdots & j_n \end{pmatrix}$$

then $\sigma^{-1} = \begin{pmatrix} j_1 & j_2 & j_3 \cdots & j_n \\ 1 & 2 & 3 \ldots & n \end{pmatrix}$

Is the average of σ. since the inverse of an even or odd permutation is even or odd respectively, the sign with $b_{1_{i1}} b_{2_{i2}} \ldots b_{n_{in}}$ and are the same. Moreover, a term like that in also appear in det A. Hence each term in the expansion of det A^T is a term in the expansion of det A and conversely Hence

For example, let.

$$A = \begin{bmatrix} 1 & -1 & 2 \\ 3 & 1 & 4 \\ 0 & -2 & 5 \end{bmatrix}, \text{ then } \quad A^T = \begin{bmatrix} 1 & 3 & 0 \\ -1 & 1 & -2 \\ 2 & 4 & 5 \end{bmatrix}$$

And $\det A = 16 = \det A^T$

Remark. The expansion of det A as given in the definition is called the expansion of determination by rows expansion of det A. Similarly, one has the notation of column expansion of det A. Show that three expansions are algebraically identical.

Remark. Also explain the fact that, in every theorem about determination, the role of columns and row may be interchange to obtain another theorem Thus, which describe the properties of det A in terms of column of A can be restarted in terms of its rows. Thus, we have

Theorem. Let A be $n \times n$ matrix and let det A denote the determination function. Then

(i) if the row vector $a_i^t = [a_{i1}\ a_{i2} \ldots a_{in}] of\ A$ is of the form $a_i^T = \alpha b_i^T$, Then $\det A = \alpha \det B$

Where B is the matrix obtained from A by replacing its ith row $a_i^T \; by \, b_i^T$

(ii) if a row $a_i^T \; of \; A \, is \, 0, then$

$\det A = 0$

(iii) if a row a_i^T of A is of form

$a_i^T = b_i^T + c_i^T$

Then

$\det A = \det B + \det C$

where B and C matrices obtain from A by replacing to ith row by b_i^T and c_i^T respectively .

(iv) if B is the matrix obtained from A by replacing its $ith \, a_i^T \; by \, a_i^T + c\, a_i^T then.$

$det\ B = det\ A.$

(v) interchanging of any two adjacent rows of A changes the sign of $det\ A$.

(vi) interchanging of any two rows of A results in the change of sign of det A.

(vii) if any two rows of A are identical then det $A = 0$

(viii) A constant multiple of a row of A may be added to any other row without changing the value of det A. That is

$$det\ A = \det\left[a_1^T \;\; a_2^t \; \ldots \; ,a_i^T \; \ldots \; a_j^T \; \ldots \; a_n^T\right] = \det\left[a_1^T \;\; a_2^T \; \ldots \; ,a_i^T + c\, a_j^T \ldots a_j^T \ldots \; a_n^T\right]$$

For any scalar c.

Proof. This theorem can be proved on the same lines as and is left as an exercise.

The main use of the above theorem lies in the fact that it facilities the evaluation of det A by its rows.

Recall that, for $n \times n$ matrix A, the following three types of operation are called elementary row (column) operations.

interchanging any two rows (column) of A.

Multiplying every element of a row (column) of A by nonzero scalar Adding a scalar multiple of a row (column) of A to any other row (column).

Also an elementary matrix is a matrix E, obtained from the identity matrix I_n by any elementary row (column) operation.

Thus

$I_{pq} =$ the matrix from I_n by interchanging the pth and qth row (column)

$I_{cp} =$ the matrix obtained from I_n by multiplying its pth row (column) by a nonzero scalar c,

I_{p+cq} = the matrix obtained from I_n by adding to the pth row (column) $c - times$ the qth row (column)

All of these are examples of elementary matrices

It is easy to see that

$del\, I_{pq} = -1$, $\det I_{cp} = c$, $\det I_{p+cq} = 1$

Also for any two elementary matrices $E_1 \; and \; E_2$.

$\det(E_1 E_2) = \det E_1 \; \det E_2$

By actual computation. For example:

$\det\left(I_{pq} I_{cp}\right) = -c = \det I_{pq} \; . \det I_{cp}$

$\det\left(I_{pq} .I_{p+xq}\right) = -1 - \det I_{pq} . \det I_{p+xq}$

With the above remarks, we are known in a position to prove the following important result about of determinates.

AN ALGORITHM TO EVALUATE DETA

By an algorithm we mean a sequence of a finite number of steps to get a desired result. The word algorithm comes from the famous Muslim mathematician Al-Khwarizmi who invented the world ALGEBRA.

A step by step evaluation of det A a= of order n is obtained as follows

Step1: By an interchange of rows of A (and taking into account the resulting sign) bring a nonzero entry to the (1,1) position (unless all the entries in the first column are zero (in which case det A=0).

Step 2. By adding suitable multiples of the first row to all other row, reduce the $(n-1)$ entries, except (1,1) in the first column , to 0 Expand det *A by its first column* repeat this process

Or continues the following steps

Step 3. Repeat step 1 and step 2 with the last remaining row concentrating on the second column.

Step 4. Repeat step 1, step 2 and step 3 with the last remaining $n-2$ rows, $n-3$ *rows* and so on, until a triangular matrix is obtained

Step 5. Multiply all the diagonal entries of the resulting triangular matrix and them multiply it by its sign to get det A.

Example. Evaluate det A, where

$$A = \begin{bmatrix} 1 & 2 & 3 & 4 \\ 2 & 3 & 4 & 1 \\ 3 & 4 & 1 & 2 \\ 4 & 1 & 2 & 3 \end{bmatrix}$$

Solution.

$$\det A = \begin{bmatrix} 1 & 2 & 3 & 4 \\ 2 & 3 & 4 & 1 \\ 3 & 4 & 1 & 2 \\ 4 & 1 & 2 & 3 \end{bmatrix}$$

$$= \begin{bmatrix} 1 & 2 & 3 & 4 \\ 0 & -1 & -2 & -7 \\ 0 & -2 & -8 & -10 \\ 0 & -7 & -10 & -13 \end{bmatrix}$$

$by\, R_2 + (-2)R_1 \quad R_3 + (-3)R_1 \; and\; R_4 +)(-4)R_1$

$$= \begin{vmatrix} -1 & -2 & -7 \\ -2 & -8 & -10 \\ -7 & -10 & -13 \end{vmatrix},$$ expanding by the first column

$$= -2\begin{vmatrix} 1 & 2 & 7 \\ 1 & 4 & 5 \\ 7 & 10 & 13 \end{vmatrix}$$ $taking(-1),(-2)\,and\,(-1)$ *common from 1st, 2nd and 3rd rows respectively*

$$= -2\begin{vmatrix} 1 & 2 & 7 \\ 0 & 2 & -2 \\ 0 & -4 & 13 \end{vmatrix}$$ by $R_2 + (-1)R_1\; R_3 + (-7)R_1$

$$= -2\begin{vmatrix} 2 & -2 \\ -4 & -36 \end{vmatrix}$$ expanding by first column

$$= -2(2)(-4)\begin{vmatrix} 1 & -1 \\ 1 & 9 \end{vmatrix}$$ *taking out 2 and -4 from 1snd 2nd row respectively*

$$= 16\begin{vmatrix} 1 & -1 \\ 0 & 10 \end{vmatrix}$$ *by* $R_2 + (-1)R_1$

$= 160.$

Example 5. Let A and B be matrices of order 6 such that $\det(AB^2) = 72,\quad \det(A^2B^2) = 144.$ Find $\det(A)\det(2A)\, and\, det$ (AB^6)

Solution. Suppose $\det A = a\, and\, \det B = b.$

Then, by the product rule if determinants, we have

$\det(BA^2) = \det A \det B^2 = \det A(\det B)^2 = ab^2 = 72$

$\det(A^2B^2) = \det A^2 . \det B^2$

$= a^2\, b^2 = 144$

So that $\det A = a = \dfrac{a^2b^2}{ab^2} = \dfrac{144}{72} = 2$

Also

$$\det(2A) = 2^6 \det A = 128.$$

Lastly,

$$\det\left(AB^6\right) = \det A \det B$$

$$= \det A \left(\det B\right)^6$$

$$= ab^6$$

$$Since \quad b^2 = \frac{72}{a} = \frac{72}{2} = 36$$

$$\det(AB)^6 = 2x(36)^3 = 93312$$

Example 6. Evaluate the determinant of the matrix

$$D = \begin{bmatrix} a & h & g \\ h & b & f \\ g & f & c \end{bmatrix}$$

Solution. Expanding by the first row, we have

$$\det D = a\,A + h\,H + gG$$

Where

$$A = \begin{vmatrix} b & f \\ f & c \end{vmatrix} = bc - f^2$$

$$H = -\begin{vmatrix} h & f \\ g & c \end{vmatrix} = fg - ch,$$

$$G = \begin{vmatrix} h & b \\ g & f \end{vmatrix} = hf - bg$$

Hence $\det D = a\left(bc - f^2\right) + h\left(fg - ch\right) + g\left(hf - bg\right)$

$$= abc - af^2 + fgh - ch^2 + fgh - bg^2$$

$$= abc + 2fgh - af^2 - bg^2 - ch^2$$

Example. Evaluate the determinant of the matrix

$$A = \begin{bmatrix} 4 & 2 & 5 & 10 \\ 1 & 1 & 6 & 3 \\ 7 & 3 & 0 & 5 \\ 0 & 2 & 5 & 8 \end{bmatrix}$$

Solution.

The solution given below makes use of and the algorithm by reducing the determinant into a determinant of smaller size. Thus

$$\det A = \begin{vmatrix} 4 & 2 & 5 & 10 \\ 1 & 1 & 6 & 3 \\ 7 & 3 & 0 & 5 \\ 0 & 2 & 5 & 8 \end{vmatrix}$$

$$= -\begin{vmatrix} 1 & 1 & 6 & 3 \\ 4 & 2 & 5 & 10 \\ 7 & 3 & 0 & 5 \\ 0 & 2 & 5 & 8 \end{vmatrix} \quad by\, R_{12}, changing\, the\, sign$$

$$= -\begin{vmatrix} 1 & 1 & 6 & 3 \\ 0 & -2 & -19 & -2 \\ 0 & -4 & -42 & -16 \\ 0 & 2 & 5 & 8 \end{vmatrix} \quad by\, R_2 - 4R_1, R_3 = 7R_1 \begin{pmatrix} The\, value \\ does\, not\, change\, with \\ these\, row\, operations \end{pmatrix}$$

$$= -\begin{vmatrix} -2 & -19 & -2 \\ -4 & -42 & -16 \\ 2 & 5 & 8 \end{vmatrix}, \quad evaluating\, by\, the\, first\, column$$

$$= -\begin{vmatrix} 2 & 19 & 2 \\ 4 & 42 & 16 \\ 2 & 5 & 8 \end{vmatrix}, \quad by\,(-1)R_1, (-1)R_2\; i.e. multiplying\, the\, determinant\, by\, (-1)^2$$

$$= -2\begin{vmatrix} 2 & 19 & 2 \\ 2 & 21 & 8 \\ 2 & 5 & 8 \end{vmatrix}, \quad taiking\, out\, 2\, from\, R_2$$

$$= -2\begin{vmatrix} 2 & 19 & 2 \\ 2 & 2 & 6 \\ 2 & -14 & 6 \end{vmatrix} \quad by\, R_2 - R_1\, and\, R_3 - R_1$$

$$= -4\begin{vmatrix} 2 & 6 \\ -14 & 6 \end{vmatrix} = -4(12+84) = -384$$

Determinant and Inverse of Matrices

The section deals with another of finding the inverse of a matrix. In this process, we shall the adjugated or adjoint of a matrix. We begin with the following simple result .

Theorem. For an invertible matrix A, $\det A \neq 0$ and

$$\det\left(A^{-1}\right) = \frac{1}{\det A}$$

Proof. Let A be a nonsingular matrix with A^{-1} as its inverse.

Then

$$A^{-1}A = 1$$

$$\det\left(A^{-1}A\right). = \det\left(A^{-1}\right)$$

$$But \quad \det\left(A^{-1}A\right) = \quad \det(I) = 1$$

$$Hence \quad \det\left(A^{-1}\right)\ \det A = 1$$

$$\det\left(A^{-1}\right) \neq 0,\ \det A \neq 0$$

$$and \quad \det\left(A^{-1}\right) = \frac{1}{detA}.$$

Before we describe the other method of finding the inverse of a matrix A, we define the adjugate or adjoint of A as follows:

Definition. Let $A = \left[a_{ij}\right]$ be an $n x n$ matrix and A_{ij} be the cofactor of a_{ij} in A.

$$B = \begin{bmatrix} A_{11} & A_{12} & \cdots & A_{1n} \\ A_{21} & A_{22} & \cdots & A_{2n} \\ \vdots & \vdots & \cdots & \vdots \\ A_{n1} & A_{n2} & \cdots & A_{nn} \end{bmatrix} be\,the\,matrix\,of\,cofactors.$$

Then the adjugated or adjoint of A, written adj A, is the transpose of B. Thus

$$adj\ A = \begin{bmatrix} A_{11} & A_{12} & \cdots & A_{n1} \\ A_{21} & A_{22} & \cdots & A_{n2} \\ \vdots & \vdots & \cdots & \vdots \\ A_{1n} & A_{2n} & \cdots & A_{nn} \end{bmatrix} = B^T$$

Example. Let

$A = \begin{bmatrix} a & b \\ c & d \end{bmatrix}$. Find $adj\ A$.

Solution. Here, the cofactors are

$$A_{11} = d, \quad A_{12} = -c, \; A_{21} = -b, \; A_{22} = a.$$

Hence $B = \begin{bmatrix} A_{11} & A_{12} \\ A_{21} & A_{22} \end{bmatrix} = \begin{bmatrix} d & -c \\ -b & a \end{bmatrix}$

so that

$$adj\ A = B^T = \begin{bmatrix} d & -b \\ -c & a \end{bmatrix}$$

Example. Let

$$A = \begin{bmatrix} 1 & 2 & 3 \\ 2 & 3 & 1 \\ 3 & 1 & 2 \end{bmatrix}. Compute\,adj\ A.$$

Solution. Here, the cofactors are

$$A_{11} = 5, \quad A_{12} = -1, \quad A_{13} = -7, \quad A_{21} = -1, \quad A_{22} = -7 \quad A_{23} = 5,$$

$A_{31} = -7, \quad A_{32} = 5, \; A_{33} = -1$

Hence, the matrix of cofactor is

$$B = \begin{bmatrix} 5 & -1 & -7 \\ -1 & -7 & 5 \\ -7 & 5 & -1 \end{bmatrix}$$

$$adj\, A = B^T = \begin{bmatrix} 5 & -1 & -7 \\ -1 & -7 & 5 \\ -7 & 5 & -1 \end{bmatrix}$$

And

Example. Let

$$A = \begin{bmatrix} 2 & 3 & 4 \\ 1 & 2 & 0 \\ 3 & 1 & 5 \end{bmatrix}$$

Examine whether A is invertible and, if so, determine A^{-1}.

Solution. We have

$$\det A = \begin{vmatrix} 2 & 3 & 4 \\ 1 & 2 & 0 \\ 3 & 1 & 5 \end{vmatrix} = -\begin{vmatrix} 1 & 2 & 0 \\ 2 & 3 & 4 \\ 3 & 1 & 5 \end{vmatrix} by\, R_{12}$$

$$= -\begin{vmatrix} 1 & 2 & 0 \\ 0 & -1 & 4 \\ 3 & -5 & 5 \end{vmatrix} by\, R_2\, R_1\, and\, R_3 - R_1$$

$= -15 \neq 0.$

Hence A is invertible.

Now, $A_{11} = 10, \quad A_{12} = -5 \quad A_{13} = -5,$

$A_{21} = -11, \quad A_{22} = -2, \quad A_{23} = 7,$

$A_{31} = -8, \quad A_{32} = 4, \quad A_{33} = 1$

Therefore,

$$A^{-1} = \frac{1}{detA} adj\ A = -\frac{1}{15}\begin{bmatrix} 10 & -11 & -8 \\ -5 & -2 & 4 \\ -5 & 7 & 1 \end{bmatrix}$$

Note. If would by interesting for the reader to compare this method this method of finding the inverse of a matrix with the one using elementary row operations on A and the identity matrix I simultaneously discussed earlier.

SUMMARY AND DISCUSSION

In this chapter we discussed about concept of determinants still of a square matrix, minors role in finding solution of linear problems. This concept played a basic role in the solution of systems of linear equations before the use of matrices and computers. The use of determinants is becoming less and less with every passing day. Nevertheless determinants still play an important but minor in finding solution of linear problems.

This chapter includes concept of determinant of a square matrix, minors and cofactors, properties of determinants, Laplace expansion of a determinant, ad joint and inverse of a matrix .

Determinant. The scalar $a_{11}a_{22} - a_{21}a_{12}$ is uniquely determined by the matrix A. It is called the **determinants of order 2** of the square matrix A given in and is denoted by det A or $|A|$. Thus.

$$|A| = \det A = a_{11}a_{22} - a_{21}a_{12} = \begin{vmatrix} a_{11} & a_{12} \\ a_{21} & a_{22} \end{vmatrix}$$

Properties

The following properties of determinants of order 2 follow directly from the above definition.

(i) For any real number k,

$$\begin{vmatrix} k\,a_{11} & a_{12} \\ k\,a_{21} & a_{22} \end{vmatrix} = \text{k} \begin{vmatrix} a_{11} & a_{12} \\ a_{21} & a_{22} \end{vmatrix}$$

(ii) if $a_2 = b_{12} + c_{12}.\ a_{22} = b_{22} + c_{22}, then$

$$\begin{vmatrix} a_{11} & a_{12} \\ a_{21} & a_{22} \end{vmatrix} = \begin{vmatrix} a_{11} & b_{12} + c_{12} \\ a_{21} & b_{22} + c_{22} \end{vmatrix}$$

$$= \begin{vmatrix} a_{11} & a_{12} \\ a_{21} & a_{22} \end{vmatrix} + \begin{vmatrix} a_{11} & c_{12} \\ a_{21} & a_{22} \end{vmatrix}$$

(iii) if the two columns of $A are$ idenitical the $\det A = 0. That$ is

$$\begin{vmatrix} a_{11} & a_{12} \\ a_{21} & a_{22} \end{vmatrix} = 0$$

(iv) The determinants of the unit matrix is 1. That is

$$\begin{vmatrix} 1 & 0 \\ 0 & 1 \end{vmatrix} = 1$$

Determinants of Order n. Firstly we define the determinant of an nxn matrix inductively, That is, from our knowledge of a determinant of order 2, we define a determinant of order 3 and use this definition of a determinant of order 3 to describe a determinant of order 4 and so on.

Thus a determinant of order 3 is defined as follows:

Let

$$A = \begin{bmatrix} a_{11} & a_{12} & a_{13} \\ a_{21} & a_{22} & a_{23} \\ a_{31} & a_{32} & a_{33} \end{bmatrix}$$

Then

$$\det A = |A| = a_{11} \begin{vmatrix} a_{22} & a_{23} \\ a_{32} & a_{33} \end{vmatrix} - a_{12} \begin{vmatrix} a_{21} & a_{23} \\ a_{31} & a_{33} \end{vmatrix} + a_{13} \begin{vmatrix} a_{21} & a_{23} \\ a_{31} & a_{32} \end{vmatrix} \quad (1)$$

Let A be a square matrix of order n. The matrix obtained from A by deleting its ith row and jth column is again a matrix M_{ij} of order $n-1. M_{ij}$ is called the ***ij th*** minor of A.

Definition. Let M_{ij} be the $ijth$ minor of a square matrix A of order n. Then $A_{ij} = (-1)^{i+j} \det M_{ij}$ is called the $ijth$ cofactor of A.

Definition. Let $A=\left[a_{ij}\right]$ be an $n\times n$ triangular matrix. Then $\det A$ is the product of the elements on the main diagonal, that is, $\det A = a_{11}a_{22}\,a_{33}\ldots a_{nn}$.

Definition. A function $\det M_n \to F\; which$ associates with a matrix A in M_n a unique element det $A\,of\,F$, is said to be the determinant function if it satisfies the following conditions:

(i) det is a linear function of column of A. That is, if

$$a_j=\beta b_j+\gamma c_j, for\ \beta,\gamma\in F,\quad 1\le j\le n,\ then$$

$$detA=\det\left[a_1\ \ a_2\ \ \ldots\ \ a_j\ \ \ldots\ \ a_n\right]$$

$$=\det\left[a_1\ \ a_2\ldots\ \ \beta b_j+\gamma c_j\ \ldots\ a_n\right]$$

$$=\beta\det\left[a_1\ \ a_2\ \ldots\ b_j\ \ \ldots\ a_n\right]$$

$$+\gamma\det\left[a_1\ \ a_2\ \ \ldots\ \ c_j\ \ \ldots\ \ a_n\right].$$

(ii) if any two adjacent columns of A are identical then $det\,A=0$

(ii) $det\,I_n=1.Here\,I_n\,is\ the\,n\,x\ n$ identify matrix and 1 is the multiplicative identity of F.

Definition. The element $\det A\,of\,F$ associated with a matrix $A.\in M_n$ is called the determinants of A.

Theorem. Let A be a square matrix of order n. Then the determinant function has the following properties.

(i) if the column vector $\boldsymbol{a}_i$ of A is of the form $a_i=a\,\boldsymbol{b}_i$. Then $\det A=\alpha\det B,\left(\alpha\,being\ a\,scalar\right)$, where B is the matrix obtained from by replacing its ith column $a_i\ by\,b_i$.

(ii) if a column $a_i\ of\ A\,is\,0,$ then $\det A=0$

(iii) if a column $a_i\,of\ A$ is of the form $a_i=b_i+c_i$, then $\det A=\det B+\det C$

where B and C are matrices obtained from A by replacing its ith column by b_i and c_i respectively

(iv) if $\boldsymbol{a}_i$ is replaced by $a_i + c\,a_{i+1}$ in A, then $\det A$ remains unchanged. That is

$$= \det[\boldsymbol{a}_1 \;\; \boldsymbol{a}_2 \ldots \;\; \boldsymbol{a}_1 + \boldsymbol{c}\,\boldsymbol{a}_{i+1} \;\; \boldsymbol{a}_{i+1} \ldots \boldsymbol{a}_n]$$

$$= \det[\boldsymbol{a}_1 \;\; \boldsymbol{a}_2 \;\ldots\;\; \boldsymbol{a}_1\,\boldsymbol{a}_{i+1} \;\ldots\; a_n]$$

(v) interchanging of any two adjacent columns of $A\,change$ the sign of $det \;\; A$

(vi) Interchanging of any two column of A results in the change of sign of det A

(vii) if any two columns of A results in the change of sign of det A.

(Viii) A constant multiplied of column of A may be added to any other column without chainging the value of $\det A, That\,is$

$$det = [a_1 \;\; a_2 \;\ldots\; a_1 + ca_j \ldots \;\; a_n],$$ for any scalar c

$$= \det[a_1 \; a_2 \;\ldots\; a_i \ldots \; a_n]$$

Theorem. Let A be $n \times n$ matrix and let det A denote the determination function. Then

(i) if the row vector $a_i^t = [a_{i1} \; a_{i2} \ldots a_{in}] of \; A$ is of the form

$a_i^T = \alpha b_i^T$, Then $\det A = \alpha \det B$

where B is the matrix obtained from A by replacing its ith row $a_i^T \; by \, b_i^T$

(ii) if a row $a_i^T \; of \; A \, is \, 0, then$

$\det A = 0$

(iii) if a row a_i^T of A is of form

$a_i^T = b_i^T + c_i^T$

Then $\det A = \det B + \det C$

where B and C matrices obtain from A by replacing to ith row by b_i^T and c_i^T respectively .

(iv) if B is the matrix obtained from A by replacing its $ith\, a_i^T$ $by\, a_i^T + c\, a_i^T then.$

$$det\, B = det\, A.$$

(v) interchanging of any two adjacent rows of A changes the sign of $det\ A$.

(vi) interchanging of any two rows of A results in the change of sign of det A.

(vii) if any two rows of A are identical then det $A = 0$

(viii) A constant multiple of a row of A may be added to any other row without changing the value of det A. That is

$$det\ A = \det\left[a_1^T \ \ a_2^t \ \dots \ ,a_i^T \ \dots \ a_j^T \ \dots \ a_n^T \right]$$

$$= \det\left[a_1^T \ \ a_2^T \ \dots \ ,a_i^T + c\, a_j^T \dots a_j^T \dots \ a_n^T \right]$$

For any scalar c.

By an algorithm we mean a sequence of a finite number of steps to get a desired result. The word algorithm comes from the famous Muslim mathematician Al-Khwarizmi who invented the world ALGEBRA.

Determinant and Inverse of Matrices

The section deals with another of finding the inverse of a matrix. In this process we shall the adjugated or adjoint of a matrix .We begin with the following simple result.

Theorem. For an invertible matrix A, $\det A \neq 0$ and

$$\det\left(A^{-1}\right) = \frac{1}{\det A}$$

CHAPTER 4

VECTOR SPACE

INTRODUCTION

Vector space from a very important topic in both pure and applied mathematics. Essentially it is a part of Linear Algebra In this chapter we discuss it and some of the concept replaced to it. Mainly we give a brief account of subspaces, their spanning sets; linear dependence and linear independence of subsets of a vector space, basic and dimension of a vector space, roe and column spaced, solution space of a linear system of equations, linear transformations and matrix of a linear transformation.

Definition and Example

Definition. Let f be a field and V a nonempty set on whose elements an operation of addition is defined. Suppose that for every $a \in F$ and every $v \in V, av\, is\, an\ element\, of\, V$. Then V is called a **vector space** over F if the following axioms hold:

(i) $V\, is\, an\, abelian\, group\ under\, addition$

(ii) $a(v+w)=av+aw$

(iii) $(a+b)v=av+bv$

(iv) $a(bv)=(ab)v$

(v) $Iv=v,1$ being the multiplicative identity of F

for all $a,b \in F; v,w \in V.$

The expression $av\ , a \in F,\ v \in V$ is called the scalar multiplicative of v by a. The elements of F are called scalars and the elements of V are called vectors. The vector space V over the frild F is sometimes written as $V(F)$. *If the feild* F, *over which* V *is a* vector space, is clearly understood form the context, then we simply write V instead of $V(F)$.

It may be noted that the scalar multiplication for a vector space V over the field F is a function $f : FV \to V$ defined by $f(a,v) = av\ , for\ all\ a \in F, v \in V.$

The additive identity of V – the zero vector – will be denoted by the hold faced 0 and the additive identity of F – the zero vector – will be denoted by 0.

A vector space V is called a real or complex vector space addition and field of real or complex numbers respectively.

Example. The set $R^3 = \{(x,y,z) : x,y\ ,z \in R\}$ is a vector space under addition and scalar multiplicative defined by :

(i) $u + u' = (x\ ,y\ ,z) + (z', y', z') = (x+, y+y', z+z'),$

$for\quad u = (x,y,z),\ \boldsymbol{u}' = (x', y', z') = (x + x', + y + y', z + z'),$

(ii) $\boldsymbol{au} = (ax, ay, az), \quad a \in \boldsymbol{R}^3.$

An element of $\boldsymbol{R}^3$ is usually written as $xi + yj + zk$ where $i = (1,0,0), \boldsymbol{j} = (0,1,0), \boldsymbol{k} = (0,0,1)$ are the unit vectors in the directions of the axes.

Physical quantities such as force, velocity momentum etc., having both magnitude and direction are represented by (1). As such we call these quantities also as vectors.

Example. Let $(F,+,.)$ be a field By definition $(F,+)$ is an abelian group. For scalar multiplication, let $x \in F$. Then, for each $a \in F, ax \in F$ by the closure law in F and

$a(x+y) = ax + ay$ (***Distributive Law in F***)

$(a+b)x = ax + ay$ (***Distributive Law in F***)

$(ab)x = a(bx)$ (***Associative Law in F***)

For all $a, b x, y \in F$

Hence $(F,+,.)$ is a vector space (over itself)

In particular the field $R, C \, and \, Q$ of real numbers, complex numbers and of rational numbers respectively are examples of vector spaces over themselves.

Example. Let V be the set of all numbers of the form $a + b\sqrt{2}$, where $a, b, \in \boldsymbol{Q}$ clearly, V is an additive abelian group. The identity of $V \, is \, 0 + 0\sqrt{2}$ and the additive inverse of $a + b\sqrt{2}$ is $-a - b\sqrt{2}$ *.if* $r \in \boldsymbol{Q}, the$ n r $\left(a + b\sqrt{2}\right) = ra + rb\sqrt{2}$ is in V Thus V is closed under scalar multiplication defined for members of Q. It is easy to verify that V satisfies the axioms (i) to (v) of definition. Hence V is a vector space over Q.

Example. Let F be a field consider the set

$F^n = \{(x_1, x_2, \ldots, x_n) : x_i \in F, 1 \le i \le n\}.$

Define addition scalar multiplication in F^n as follows:

For $(x_1, x_2, \ldots, x_n)$, $y = (y_1, y_2, \ldots, y_n) \in F^n$,

We define

$x + y = (x_1 + y_1, x_2 + y_2, \ldots x_n + y_n)$

And for $a \in F$, we put

$$ax = (ax_1, ax_2, \ldots, ax_n)$$

Then $(F^N, +)$ is an abelian group under addition define by (1) with $0 = (0, 0 \ldots, 0)$ as the additive identity and for each $x \in F^n$, $-x = (-x_1 - x_2, \ldots, -x_n)$ is the additive inverse.

Also $a(x+y) = a(x_1 + y_1, x_2 + y_2, \ldots, x_n + y_n)$

$$= (a(x_1 + y_1), a(x_2 + y_2), \ldots, a(x_n + y_n), by(2)$$

$$(ax_1 + ay_1, ax_2 + ay_2, \ldots ax_n + ay_n),$$

By Distributive Law in F

$$= (ax_1, ax_2, \ldots, ax_n) + (ay_1, ay_2, \ldots, y_n)$$

$$= a(x_1, x_2 \ldots, x_n) + a(y_1, y_2, \ldots, y_n),$$

$$= ax + ay$$

For all $a \in F \; and \; x, y \in F^n$ so is satisfied

The condition can be verified similarly

As a special case when $F = \boldsymbol{R}$, then we write R^n for F^n. So $\boldsymbol{R}''$ is a vector space for each $n = 1, 2, \ldots$ We call $\boldsymbol{R}'' \; the$ n-dimensional Euclidean space

We usually write vector in $\boldsymbol{R}''$ in the form of row vectors, $if \; x \in \boldsymbol{R}''$, we write $x = (x_1, x_2, \ldots, x_n) \; where \; x_1, x_2 \ldots, x_n$ belong $\boldsymbol{R}$.

Example. Let M_{mn} denote the set of all $m x n$ matrices with entries from the field of real numbers. Then M_{mn} is a vector space R under the operations of matrix addition and scalar multiplication of matrices by real numbers. The $m x n$ zero matrix is the zero of M_{mn}.

Example. Recall that a polynomial of degree n over a field $F,$ in a variable $x,$ is an expression of the form

$$p(x)=a_0+a_1x+a_2x^n, \quad a_i\in F \tag{1}$$

Consider the set $P(x)$ of all polynomial of arbitrary degree. Elements of $P(x)$ are of the from For $P(x),q(x)\in P(x),$

$$q(x)=b_0+b_1x+b_2x^2+\ldots+b_mx^m, b_i\in F,$$

We define addition and scalar multiplication in $P(x)by:$

$$p(x)+q(x)=(a_0+b_0)+(a_1+b_1)x+(a_2+b_2)x^2+(a^m+b^m)x^m+a_{m-1}x^{m-1}+\ldots+a_nx^n, (m<n) \tag{2}$$

And

$$kp(x)=(ka_0)+(ka_1)x+(ka_2)x_2+\ldots+(ka_n)x^n \tag{3}$$

The zero polynomial, which is assigned no degree, I the additive identity of $P(x).$ Then, under the addition and scalar multiplication defined by respectively, $O(x)is$ a vector space over $F.$

In particular, the set $P_n(x)$ of all polynomials of degree $\leq n,n$ fixed, together with the zero polynomial, is also a vector space over $F.$

Example. Let X be a nonempty subset of **R**. Let V be the set of all functions $f:X\to R.$ We define addition and scalar multiplication in V as follows:

For $f,g\in V\ put$

$$(f+g)(x)=f(x)+g(x) \tag{1}$$

And $(af)(x)=a.f(x)$ (2)

For all $x\in X\ and\ \in \boldsymbol{R}.$

(Note that for each $x\in X\ and\ a\in \boldsymbol{R}.$

We show that V is a vector space over $\boldsymbol{R}.$

$Let\ f,g,h\in \boldsymbol{V}.$ Then

$((f+g)+h(x))=(f+g)(x)+h(x),$ by (1)

$=(f(x)+g(x))+h(x),$ by (1)

$=f(x)(g(x)+h(x)),$ by Associative Law in **R.**

$=f(x)+g+h)(x)$ by (1)

$=(f+(g+h)),(x)$ by (1)

For all $x \in X.Hence, by$ the equality of two functions, we have

$(f+g)+h=f(g+h),$ $\quad for\,all\; f,g,h \in V$

The zero function $0:\rightarrow \boldsymbol{R}$ defined by:

$0(x)=0 \quad for\,all\; x \in X$ (3)

Is the additive identity in V. Here, for any $f \in \boldsymbol{V},$

$(\boldsymbol{f}+0)(\boldsymbol{x})=\boldsymbol{f}(\boldsymbol{x})+0(\boldsymbol{x}),$ by (1)

$=f(x)+0,$ by (3)

$=f(x)$ since 0 is the additive identity of **R**

So $f+0=f$ $\quad for\,all\; f \in V.$

Similarly, $0+f=f.$

Next, $for\,each\; f \in V, the\; function(-f):X \rightarrow \boldsymbol{R} defined\; by:$

$(-f)(x)=-f,(x),$ for all $x \in X$

Is the additive inverse of $f in V$. Here

$(f+(-f))(x)=f(x)+(-f)(x),$

$=f(x)-f(x),$

$=0$

$= 0(x)$

For all $x \in x$ Hence

$f + (-f) = 0$

Similarly, $(-f) + f = 0$

Moreover, for any $f, g \in V$,

$(f + g)(x) = f(x) + g(x)$,

$= g(x) + f(x)$ (Commutative Law in R)

$= (g + f)(x)$,

For all $x \in X$ *Hence*

$f + g = g + f$

So $(V, +)$ is an abelian group.

(ii) for the other axioms of a vector space for $v, Let\, F, g \in V\ , a \in \boldsymbol{R}$

Then, for all $x \in \boldsymbol{X}$,

$(a(f + g))(x) = a(f + g)(x)$,

$= a(f(x) + g(x))$,

$= af(x) + ag(x)$

$= (af)(x) + (ag)(x)$

$= (f + ag)(x)$

And, therefore, $a(f + g) = af + ag$

(iii) For $a, b \in \boldsymbol{R}, f \in V$ and all $x \in X$,

$((a + b)f)(x) = (a + b)f(x)$,

$= af(x) + bf(x)$

$= (af + bf)(x)$

Thus $(a+b)f = af + bf.$

(iv) Again, $(a(bf))(x) = a(bf(x)),$

$= ab\,f(x)$

$= (ab)f(x)$

$= ((ab)f)$

For all $x \in \boldsymbol{X}$, *showing that* $a(bf) = (ab)f.$

(v) $(1f)(x) - 1f(x) = f(x)$

And so $1f = f.$

Thus V is a vector space over $\boldsymbol{R}$.

Theorem. Let V be a vector space over a field F. then

(i) $a0 = 0,$ *for all* $a \in F$

(ii) $0v = 0,$ *for all* $v \in V$

(iii) $(-a)v = a(-v) = -av,$ *for all* $a \in F, v \in V$

(iv) *if* $av = 0$ *then either* $a = 0$ *or* $v = 0$ *for all* $a \in F$, $\mathbf{v} \in V$

(v) $a(u - v) = au - av,$ *for all* $a \in F, u, v \in V.$

Proof. The equation $a0 = 0$ should be read as "a (zero vector) = 0 *vector*" and likewise $0v = 0$ should be read as "$(scalar)v = 0\,vector$ "

(i) Since 0 is the additive identity of V, we have

$a0 = a(0 + 0)$

$= a0 + a\theta,$

Since $a0 + 0 = a\theta$, we have from $a0 = 0$

(ii) In this case

$0 = 0v,$

$= (a + (-a)v.$

$= av + (-a)v,$

Adding $-av$ to both sides, and using the fact that $-av + av = 0.$ we obtain

$= -av(-a)v$

Again, $0 = a0$

$= a(v + (-v))$

$= av + a(-v),$

As before, we get

$-av = a(-v)$

$(-av) = a(-v) = -av$

(iv) Let $av = 0$ with $a \neq 0.$ Then there exist an element $a^{-1} \in F$ such that $aa^{-1} = 1$

Now

$v = 1v$,

$= (aa^{-1})v$

$= a^{-1}(av),$

$= a^{-1}0 = 0$

Showing that v is zero vector

Next, let $av = 0$ but v $\neq 0.$ We have to show that $a = 0.$

Suppose $a \neq 0.$ Then a^{-1} exist and

$0 = a^{-1}(0) = a^{-1}(av) = (aa^{-1})v = 1v = v$

Contradicting that $v \neq 0.$ Thus

$a = 0$.

(v) $Let\, u, \in V, a \in F$. Then

$a(u - v) = a(u + (-v))$

$= a(u + a(-v))$

$= au + (-av),$

$= au - av,$

SUBSPACES

Definition. A nonempty subset W of a vector space V over a field F is said to be a subspace of V if W is itself a vector space over F under the same operations as defined in V.

Every vector space V has two subspaces, namely V Any other subspace of V is called a nontrivial subspace.

Example. The space $P_n(x)$ of all polynomial degree $\leq n,$ together with the zero polynomial, is a subspace of the space $P(x)$ of all polynomials.

Example. $Let\, V$ be the Euclidean space R^3. The set W consisting of those vector whose third coordinate is zero, $i.e,$ vector of the form $(x, y, 0); x, y \in \boldsymbol{R}$ is a subspace of V.

Theorem. Let V be a vector space over a field F and let W be a nonempty subset of V. Then W is a subspace of V if and only if.

$w_1, w_2 \in W \qquad \Rightarrow w_1 + w_2 \in W$

$w \in W, a \in F \qquad \Rightarrow aw \in W$

In other words W is a subspace of $V,$ if and only if, it is closed under the operations of addition and scalar multiplication as defined in V.

Proof. Suppose that W is a nonempty subset of V satisfying and then We show that W is a vector space,

$Let\, w_1, w_2 \in W. Then$ by

$(-1) w_2 \in W, i.e., -w_2 \in W. Now\, w_1, -w_2 \in W\, and\, so\, by\, , w_1, w_2 \in W.$

Therefore, W is a subgroup of the group V under addition Since V is an abelian group W is also an abelian group. Thus, axiom of is satisfied for W The axioms to of a vector space hold in W since they hold in the larger set V. Hence W is a vector space over F.

Conversely, if W is a subspace of V, then conditions and hold by definition

The two necessary and sufficient conditions of theorem for W to be a subspace of V can be combined into one, which sets in the following corollary.

Corollary. A nonempty subset W of V is a subspace of V, if and only if, $w_1, w_2 \in W\, and\, a, b \in F$ imply $aw_1 + bw_2 \in W.$

Proof. Suppose that $w_1, w_2 \in W\, and\, a, b \in F\, imply \quad aw_1 + bw_2 \in W$. Taking $a = b = 1, we\, have\, w_1 + w_2 \in W.$ Taking $b = 0,$ we get $aw_1 \in W.$ Hence, by Theorem, W is a subspace of V by Theorem W is a subspace of V .

Conversely, let W be a subspace of V. Then $aw_1 + bw_2 \in W.$ by of. Hence $aw_1 + bw_2 \in W\, by\, of\, theorem$ Thus the condition is satisfied.

Example. Consider the homogenous system, $Ax = 0\, of\, n$ linear equation in n real unknown variables. The system has a nontrivial solution x^* if and only if the matrix A is singular .SO suppose that A is a singular $n x n$ matrix. Then the system

$Ax = 0$

Has an infinite number of solution. Let S be of all such solutions, The members of S are row vector from $R^n. Thus\, S \subseteq R^n.$

If $x, y \in S, then$

$Ax = 0, \quad ay = 0$

So $A(x + y) = Ax + Ay = 0$

Hence $x+y\in S$.

Next for any $c\in R\,and\,x\in S$.

$A(cx)=cAx=c.\ 0=0$

So $cx\in S$.

Hence S is a subset of R^n

Theorem. Let U and W be subspace of a vector space V over a field F. Then $U\cap W$ is also a subspace of V.

Proof. Let $a,b\in F\,and\,w_1,w_2\in U\cap W$. Then $w_1,w_2\in U\,and\,w_1,w_2\in W$. Since U and W are subspace of V, by Corollary $aw_1\_bw_2$ belongs to both U and W. Thus $aw_1+bw_2\in U\cap W$. Hence $U\cap W$ is subspace of V.

Theorem. The intersection of any number of subspace of a vector space V is a subspace proof V.

Definition. (Sum of two space). Let $U\,and\,W$ be two subspaces of a vector space V. Define

$U+W=\{u+w:u\in U\,and\,w\in W\}$

Clearly $U+W\subset V$.

We next show that $U+W$ is a subspace of V.

Theorem. if U,W are subspace of a vector space $V,\,then\,U+W$ is a subspace of $V\,contatining\ both\,U\,and\,W$. Further, $U+W$ is the smallest subspace containing both $U\,and\,W$.

Proof. $Let\,v_1,v_2\in U+W$. Then we can write

$v_1=u_1+w_1,v_2=u_2+w_2,where\,u_1,u)2\in U\,and\,w_1,w_2\in W.$

Hence $v_1+v_2=(u_1+w_1)+(u_2+w_2)=(u_1+u_2)+(w_1+w_2)$

But since U,W are subspace so $u_1+u_2\in U\,and\,w_1,w_2\in W$. Thus $v_1+v_2\in U+W$. Also $a\mu_1\in U\,and\,aw_1\in W for$ all $a\in F$. Therefore,

$a\mu_1 + aw_1 = a(\mu_1 + w_1 \in U + W.Hence$ by theorem U + W is subspace of V.

Since $0 \in W,]mu + 0 \in U + W \; for\, every \; \mu \in U$. Thus $U \subset U + W$. similarly, $W \subset U + W$. Thus U, W are both contained in $U + W$.

Let S be any subspace of V contatining both $U\, and\, W$. Then, for every $u \in U$ and $w \in W, we\, have\, u \in S\, and\, w \in S$ so that $u + w \in S \; so\, that \; u + w \in S$ and hence $U + W \subset S$. Thus $U + W$ is the smallest subspace contatining both x $U \; and \; W$.

Definition. $Let\, U, W$ be subspace of a vector space V. if $U \cap W = \{0\}$, then Thus a vector space V is called the direct sum of its subspace $U \; and \; W$ if

(i) $v = U + W$ and

(ii) $U \cap W = \{0 \cap\}$

LINEAR COMBINATION AND SPANINING SETS

Definition. Let V be a vector space over a field F and $v_1, v_2, \ldots . v_m \in V$. Any vector in V of the form $a_1v_1 + a_2v_2 + \ldots + a_mv_m$, where $a_i \in F(1 \le i \le m)$, is called a linear combination (over F) of $v_1, v_2, \ldots v_m$.

Definition. Let S be a nonempty subset of a vector space V over a field F. The $\langle S \rangle$ is a subspace of V containing S and it is the smallest subspace of V containing S.

Proof. Let $u, v \in \langle S \rangle and\; a, b \in F$. Then there are vectors $u_1, u_2, \ldots u_m, v_1, v_2, \ldots, v_n$ in S and scalar $a_1, a_2, \ldots, a_m, b_1, b_2, \ldots, b_n \; in\, F$ such that

$$u = a_1u_1 + a_2u_2 + \ldots + a_mu_m, \; v + b_1v_1 + b_2v_2 + \ldots \; b_nv_n$$

So, for $a, b \in F$,

$$au + bv = (aa_1)u_1 + (aa_2)u_2 + \ldots + (aa_m)U_m + (bb_1)v_1 + (bb_2)u_2 + \ldots + (bb_n)v_n$$

Being a linear combination of elements of $S, is\, in \langle S \rangle$. By Corollary, $\langle S \rangle$ is a subspace of V.

Each vector $v_i\, in\, S\; is\, a$ linear combination of $v_1, v_2, \ldots, v_k$ in S since $v_i = 0v_1 + 0v_2 + \ldots + 0v_{i-1} + 1v_i + 0v_{i+1} + \ldots + 0v_K$

Thus $\langle S \rangle$ conatins each of the vectors $v_1, v_2, \ldots, v_n$ belonging to S i.e. $S \subset \langle S \rangle$.

If W is any other subspace V containig $S,$ then it contain all vectors of the from $\sum_{i=1}^{k} a_i v_i,$ where $a_i \in F, v_i \in V\, and\, K$ is natural numbers. Hence $\langle S \rangle W.$ Thus $\langle S \rangle$ is the smallest subspace of V containing $S.$ $\langle S \rangle$ is said to be spanned (or generated) by $S\, and\, S$ is called the spanning set for $\langle S \rangle$.

Corollary. $if\; S, T$ are subsets of V then
$S \subset T \quad implies \langle S \rangle \subset \langle T \rangle.$

Definition. A vector space V is said to be finite dimensional if there is a finite subset $S\, in\, V$ such that $V = \langle S \rangle.$

Example. $In V + R^3, let\, S$ be the subset consisting of vectors $e_1 = (1,0,0), e_2 = (0,1,0), e_3 = (0,0,1), i.e., S = \{e_1, e_2, e_3\}.$

Then $\langle S \rangle$ contains all vectors of the form

$$a_1 e_1 + a_2 e_2 + a_3 e_3, where\, a_1, a_3 \in R.$$

Now $a_1 e_1 + a_2 e_2 + a_3 e_3 = a_1(1,0,0) + a_2(0,1,0) + a_3(0,0,1)$
$= (a_1, a_2, a_3)$

So $\langle S \rangle \subseteq \boldsymbol{R}^3$.

But (a_1, a_2, a_3) is an arbitrary vector in $R^3 .So\, R^3 \subseteq \langle S \rangle$. Thus $\langle S \rangle = R^3$

Since S consist of a finite number of elements of R^3, by definition. $\boldsymbol{R}^3\, is$ finite dimensional (over R)

Example. Write the vector $v = (1,\ -2,\ 5) \in R^3$ as a linear combination of the vector $v_i = (1,1,1)$, $v_2 = (1,2,3)\, and = (2,-1,1)$.

Solution. We have v as

$$v = a_1 v_1 + a_2 v_2 + a_3 v_3,\quad for\, a_1\ , a_2, a_3 \in R$$

$$i.e,\ (1\ ,-2\ ,5) = a_1 v_1 + a_2 v_2 + a_3 v_3,\quad for\, a_1, a_2, a_3 \in R$$

$$\text{i.e,}\ (1,-2\ ,5) = a_1 (1,1,1) + a_2 (1,2,3) + a_3 (2,-1,1)$$

$$= (a_1 + a_2 + 2a_3, a_1 + 2a_2 - a_3, a_1 + 3a_2 + a_3)$$

Therefore, $a_1 + a_2 + 2a_3 = 1$

$$a_1 + 2a_2 - a_3 = -2$$

$$a_1 + 3a_2 + a_3 = 5$$

Solving the above system of equations by the methods of, we obtain $= a_1 = -6,\quad a_2 = 3,\ a_3 = 2.$

Hence $v = (1,-2,5) = -6v_1 + 3v_2 + 2v_3$ is the desired linear combination.

Example. Find equations defining the subspace W of R^3 spanned by the vector $(2,3,4)$.

Solution. Every vector $v = (x, y, z)\, of\, W$ is a linear combination of (2,3,4). Therefore

$$(x, y\ , z0 = t(2,3,4),\quad t \in R$$

Or $x = 2t,\ y = 3t,\ z = 4t$

Which are parametric equations of a straight line through $(0,0,0)$ with direction vector $(2,3,4)$. This straight line is the subspace by $(2,3,4)$.

Example. $Let\, V = P(x)$ be the vector space of all polynomials over a field F .

$S = \{1, x, x^2, x^3, \ldots\}, then < S \geq= V$ because any polynomial is a linear combination of 1 and powers of x. Since S is not a finite subset of V, we infer that V is not finite dimensional. It is easy to see that a finite number of elements in S cannot generate V.

A vector space which is not finite dimensional is called infinite dimensional. Thus $P(x)$ is an infinite dimensional vector space.

Example. Show that the set $S = \{(x, y, z) \in R^3 : wx + 3y - 4z = 2\}$ is not a subspace of R^3.

Solution. The point $(1\,,0,0) \in S. But\, 3(1,0,0) - (3,0,0)$ does not belong to S. Thus is not satisfied and so S is not subspace of R^3.

Linear Dependence and Basis

Definition. $Let\, V$ be a vector spave over a field F. The vector $v_1, v_2 \ldots, v_m m \in V$ are said to be linearly dependent over F if

$av_1 + a_2 v_2 + \ldots + a_m v_m = 0\, implies\ each\, a_1 = 0, i - 1, 2, \,, m.$

In this case we say that $\{v_1, v_2, \ldots, v_m\}$ is a linearly independent set.

An infinite set S of vectors is linearly independent if each of its finite subsets is linearly independent. Otherwise it is linearly dependent.

The empty set $\varnothing$ is defined to be linearly independent.

It may be noted that if one of the vectors $v_1, v_2, \ldots, v_m$ is zero say $v_i = 0,$ then the vector is linearly dependent, because

$0v_1 + 0v_2 + \ldots + 1v_i + 0v_{i+1} + \ldots + 0v_m = 0$ and the coefficient of v_i is not zero.

On the other hand, any nonzero vector v, by itself, is linearly independent, for v = 0 and $v \neq 0 \, imply \, a = 0$

Example. Show that vectors $(1,0,0),(0,1,0),(0,0,1)$ belonging to R^3 are linearly independent over ***R***.

Solution. $a(1,0,0)+B(0,1,0)+C(0,0,1)=(0,0,0)$ implies that

$(a,b,c)=(0,0,0)$

i.e, $a=0, b=0, c=0$

Thus the given vectors are linearly independent over **R.**

Example. Show that the vectors $(3,0,-3)$, $(-1,1,2),(4,2,-2) \, and \, (2,1,1)$ are linearly dependent over ***R***.

Solution. Let

$a(3,0,-3)+b(-1,1,2)+c(4,2,-2)+d(2,1,1)=(0,0,0)$

Where $a, b, c, d \in \boldsymbol{R}$

i.e, $(3a-b+4c+2d, b+2c+d, -3a+2b-2c+d)=(0,0,0)$

Therefore, $3a-b+4c+2d=0$ (1)

$b+2c+d=0$ (2)

$-3a+ab-2c+d=0$ (3)

Subtracting equation (2) from (3)

$-3a+b-4c=0$ (4)

By adding (1) and (4), we get $d=0. So (2)$ becomes

$b+2c=0. \quad That \, is \, b=-2c.$

Equation (4) and (5) with $d=0 \, give -3a=6c \;\; so \, that$

$a=b=-2c, d=0.$

So $-2(3,0,-3)-2\ (-1\ ,1,2)+1(4,2,-2)+0(2,1,1)=(0,0,0)$

Therefore the given vectors are linearly dependent.

Example. What condition must a,b,c and d satisfy so that the matrices

$$\begin{bmatrix} 1 & 2 \\ -1 & 0 \end{bmatrix}, \begin{bmatrix} 2 & 3 \\ -2 & 1 \end{bmatrix} and \begin{bmatrix} a & b \\ c & d \end{bmatrix}$$

In $m_{22}\, are\, linearly$ dependent?

Solution. Since the matrices are to linearly dependent, there must exist scalars $c_1, c_2, c_3,$ not all zero, such that

$$c_1 \begin{bmatrix} 1 & 2 \\ -1 & 0 \end{bmatrix} + c_2 \begin{bmatrix} 2 & 3 \\ -2 & 1 \end{bmatrix} + c_3 \begin{bmatrix} a & b \\ c & d \end{bmatrix} = \begin{bmatrix} 0 & 0 \\ 0 & 0 \end{bmatrix} \qquad (1)$$

This implies

$$\begin{bmatrix} c_1 + 2c_2 + ac_3 & 2c_1 + 3c_2 + bc_3 \\ -c_1 - 2c_2 + cc_3 & c_2 + dc_3 \end{bmatrix} = \begin{bmatrix} 0 & 0 \\ 0 & 0 \end{bmatrix}$$

i.e., the system of homogenous equations

$$\left.\begin{aligned} c_1 + 2c_2 + ac_3 &= 0 \\ 2c_1 + 3c_2 + bc_3 &= 0 \\ -c - 2c_2 + cc_3 &= 0 \\ c_2 + dc_3 &= 0 \end{aligned}\right\} \qquad (2)$$

must have a nontrivial solution.

By theorem the system will have a nontrivial solution if rank of the matrix

$$A = \begin{bmatrix} 1 & 2 & a \\ 2 & 3 & b \\ -1 & -2 & c \\ 0 & 1 & d \end{bmatrix}$$

is less than 3 i.e., Rank $A = 2$ at the most

Now,

$$A\tilde{R}\begin{bmatrix} 1 & 2 & a \\ 0 & -1 & b-2a \\ 0 & 0 & c+a \\ 0 & 1 & d \end{bmatrix} \text{ by } R_2 - 2R_1 \; and \; R_3 + R_1$$

$$\tilde{R}\begin{bmatrix} 1 & 2 & a \\ 0 & -1 & b-2a \\ 0 & 0 & c+a \\ 0 & 1 & b+d-2a \end{bmatrix} by\, R_4 + R_2$$

The last matrix has Rank 2 at the most only if

$c+a=0 \quad and\, b+d-2a=0$

Thus, the require conditions on a,b,c,d are '

$a+c=0 \quad and \quad b+d-2a=0.$

Thus, the required conditions on a,b,c,d are

$a+c=0 \quad and\, b+d-2a=0$

Note that, for $a=b=d=2\;,c=-1 \; and\; c_1=1, c_2=-1\;, c_3=1,$ the equation (1) satisfied.

Theorem. A Set $S=\{v_1, v_2, \ldots, v_n\} of\, n$ vectors $(n \geq 2)$ in a vector space V is linearly dependent if and only if at least one of the vectors in S is a linear combination of the remaining vectors of the set

(ii) A set $S=\{v_1, v_2, \ldots, v_n\}$ of vector $n \geq 2,$ in a vector space V is linearly dependent if and only if one of the vectors in S is a linear combination of the vectors preceding it.

Proof. (i) Suppose the set $S=\{v_1, v_2, \ldots, v_n\}$ is linearly dependent. Then there are scalars $a_1, a_2, \ldots, a_n,$ at least one of them, say $a_i\, is$ nonzero, such that

$a_1v_1 + a_2v_2 + \,..+a_iv_i + \ldots + a_nv_n = 0$

Dividing through by a_i and transposing, we have

$$v_i = -\frac{1}{a_i}\left(a_1 v_1 + a_2 v_2 + \ldots + a_{i-1} v_{i-1} + a_{i+1} v_{i+1} + \ldots + a_n v_n \right)$$

Expressing v_i as a linear combination of the remaining vectors of the set,

Conversely, let some vectors v_j of the given set be a linear combination of the remaining vectors i.e., $v_j = a_1 v_1 + a_2 v_2 + \ldots + a_{j-1} v_{j-1} - v_j + a_{j+1} v_{j+1} + \ldots + a_n v_n$.

Then $a_1 v_1 + a_2 v_2 + \ldots + a_{j-1} v_{j-1} - v_j + a_{j+1} v_{j+1} + \ldots + a_n v_n = 0$ and there is at least on coefficient, namely -1 of v_j, which is nonzero and therefore by definition, the set

$\{v_1, v_2, \ldots, v_{j-1}, v_j, \ldots, v_n\}$. is linearly dependent.

(iii) Suppose that the set $S = \{v_1, v_2, \ldots, v_n\}$ is linearly dependent. Then there are scalars $a_1, a_2, \ldots, a_n$ not all zer, such that

$$a_1 v_1 + a_2 v_2 + \ldots + a_n v_n = 0$$

Let a_k be the least nonzero scalars in. Then the terms $a_{k+1} v_{k+1}\, a_{k+2} \ldots, a_n v_n$ are all zero. So can be written as

$$a_1 v_1 + a_2 v_2 + \ldots + a_k v_k = 0, a_k \neq 0,$$

so that

$$v_k = -\frac{1}{a_k}\left(a_1 v_1 + a_2 v_2 + \ldots + a_{k-1} v_{k-1} \right)$$

Hence v_k is a linear combination of the vectors preceding it.

Conversely, suppose that in $S = \{v_1, v_2, \ldots, v_n\}\, some$ one of the vectors, v_k say, is a linear combination of the vectors preceding it. That is $v_k = b_1 v_1 + b_2 v_2 + \ldots + b_{k-1} v_k = 0,$

Then

$$b_1v_1 + b_2v_2 + \ldots + b_{k-1}v_{k-1} + (-1)v_k = 0,$$

so that

$$b_1v_1 + b_2v_2 + \ldots + b_{k-1}v_{k-1} + (-1)v_{k+0}\, v_{k+1} + \ldots + 0.v_n = 0$$

with not all coefficient in zero. Hence S is linearly dependent.

Definition. A set of linearly independent vectors spanning a vector space V is called a basis for V. Unless $V = \{0\}$, the basis vectors are nonzero, since we can ignore any zero vector from a spanning set.

Example. The vectors $e_1 = (1,0,0), e_2 = (0,1,0), e_3 = (0,0,1)\, span\, R^3$ and since these vectors are linearly independent, it follows that the set $\{e_1, e_2, e_3\}$ forms a basis of $R^3.[\{e_1, e_2, e_3\}\, is\, called$ the standard basis for R^3].

Example. Show that the 2×2 matrices

$$E_1 = \begin{bmatrix} 1 & 0 \\ 0 & 0 \end{bmatrix}, E_2 \begin{bmatrix} 0 & 1 \\ 1 & 0 \end{bmatrix}, E_3 = \begin{bmatrix} 0 & 0 \\ 0 & 1 \end{bmatrix}$$

From a basis for the vector space V of all 2×2 symmetric matrices.

Solution. Recall that 2×2 symmetric matrix is of the form

$$A = \begin{bmatrix} a & b \\ b & c \end{bmatrix}$$

$\{E_1, E_2, E_3\}$ is a basis of the vector space V of all 2×2 symmetric matrices if and only if every symmetric matrix A is linear combination of E_1, E_2, E_3. But

$$\begin{bmatrix} a & b \\ b & c \end{bmatrix} = a\begin{bmatrix} 1 & 0 \\ 0 & 0 \end{bmatrix} + b\begin{bmatrix} 0 & 1 \\ 1 & 0 \end{bmatrix} + c\begin{bmatrix} 0 & 0 \\ 0 & 1 \end{bmatrix}$$

and E_1, E_2, E_3 are linearly independent. (Verify!)

So $\{E_1, E_2, E_3\}$ is a basis of V.

Can this result he generalized to:

the space of 3×3 symmetric matrices,

the space of 4×4 symmetric matrices,

to space of $n\times n$ symmetric matrices?

Theorem. Any finite dimensional vector space contain basis.

Proof. Let $v_1,v_2,\dots,v_r$ be a finite spanning set for the vectors space V assumed not to be $\{0\}.if\ v_1,v_2,\dots,v_r$ are linearly independent, they from a basis for V and there is nothing to prove.

Suppose that $v_1,v_2,\dots,v_r$ are not linearly independent. Then these vectors are linearly dependent. Therefore by One of the vectors say v_r is a linear combination of the remaining ones We cast out this vectors from the set and, renumbering if necessary , we obtain a set of $r-1\,vectors\ v_1,v_2,\dots,v_{r-1}.$ Clearly any linear combination of $\{v_1,v_2,\dots,v_{r-1}\}$ is also a spanning set for $V.$ Continuing in this way we arriving at a linearly independent spanning set $\{v_1,v_2,\dots,v_n\}(1\le b<r)$ and this forms a basis for $V.$ Thus every finite dimensional vector space contain a basis.

Definition. The dimension of a vector space V over a field $F,$ written dim $V,$ is the number of elements in a basis for V .

In the follows from Theorem, the dimension of a vector space C cannot exceed the number of elements in a spanning set for $V.$

Theorem. All bases of a finite dimensional vector space contain the same number of elements.

Proof. Suppose that the vector space V has teo bases A and B with m and n elements respectively. Since A spans V and B is linearly independent, it follows from the previous remark that the number of elements in B cannot exceed the number of elements in $A.$ Thus $n\ge m.$ By interchanging the roles of A and $B,$ we have $m\le n.$ Hence $n=m.$

Example. Find a basis for the plane $y-z=0$ considering it as a subspace of $\boldsymbol{R}^3$.

Solution. The general solution of the equation $y-z=0 is\ y=z.$ In vectorfromthegeneralsolutionis $(x,y,z)=(x,y,z)=x(1,0,0)+y(0,1,1)$.

Thus, the plane $y-z=0$ is spanned by the vectors $(1,0,0) and (0,1,1)$. The dimension of the given subspace is two, since the vectors $(1,0,0)$ and $(0,1,1)$ from a basis for the given plane. [the vector $(1,0,0),(0,1,1)$ are linearly independent]

Theorem. Let V be a space such that dim $V=n<\infty, A$ set of vectors $\{v_1,v_2,\ldots,v_n\}\subset V$

Is a basis for V if and only if each vectors in V is uniquely expressible as a linear combination of $v_1,v_2,\ldots,v_n$.

Proof. Let the set $\{v_1,v_2,\ldots,v_n\}$ be a basis for $V\cdot$ Then every vector $v\,in V\ can be\ expressed$ in at least one way as a linear combination of the basis vector i.e.,

$$v=a_1v_1+a_2v_2+\ldots+a_nv_n \qquad (1)$$

where $a_i, 1\le n,$ are scalars.

Suppose v is also expressible as

$$v=b_1v_1+b_2v_2+\ldots+b_nv_n \qquad (2)$$

$b_1,b_2,\ldots,b_n$ being scalars.

Subtracting equation (2) from (1), we get

$$0=(a_1=b_1)+(a_2-b_2)v_2+\ldots+(a_n-b_n)v_n.$$

$$a_1-b_1=a_2-b_2=\ldots=a_n-b_n=0$$

i.e., $a_i=b_i,\quad 1\le i\le n.$

Thus every vector $v in V$ can be expressed in a unique way as a linear combination of the basis vectors.

Conversely, let every vector of V be uniquely expressible as a linear combination of $v_1, v_2 \ldots, v_n$

Then these vectors span V. We prove that they are linearly independent.

Suppose that, for scalars $a_1, a_2, \ldots, a_n$

$a_1 v_1 + a_2 v_2 + \ldots + a_n v_n = 0.$

Since $0v_1 + 0v_2 + \ldots + 0v_n = 0,$

The uniqueness of representation (1) implies $0 = a_1 = a_2 = \ldots = a_n.$

Thus $v_1, v_2, \ldots, v_n$ are linearly independent. Since $v_1, v_2, \ldots, v_n$ span V and are also linear independent, they from a basis for V.

Theorem. *Let* $S = \{v_1, v_2, \ldots, v_n\}$ be a basis for $n = dimensional$ vector space V over a field F. Then every set with more than n vectors is linearly dependent.

In order to show that the set B is linearly dependent we must find scalars $c_1, c_2 \ldots, c_r$, not all zero, such that

$c_1 u_1 + c_2 u_2 + \ldots + c_r u_r = 0.$

Since the set $S = \{v_1, v_2, \ldots, v_n\}$ is linearly independent, by each u_1 can be expressed in a unique way as a linear combination of $v_1, v_2, \ldots, v_n$. Thus

$$\left.\begin{array}{l} u_1 = a_{11} v_{11} + a_{12} v_2 + \ldots + a_{1n} v_n \\ u_2 = a_{21} v_1 + a_{22} v_2 + \ldots + a_{2n} v_n \\ \vdots \quad \vdots \quad \vdots \quad \ldots \quad \vdots \\ u_r = a_{r1} v_1 + a_{r_2} v_2 + \ldots + a_{rn} v_n \end{array}\right\}$$

Where a_{ij} are scalar $(1 \le i r, 1 \le j \le n).$

Subtracting the value of $u_i, i = 1, 2, \ldots, r,$ from (2) into (1), we have

$$c_1(a_{11}v_1 + a_{12}v_2 + \ldots + a_{1n}v_n) + c_2(a_{21}v_1 + a_{22}v_2 + \ldots + a_{2n}v_n)$$

$$+ \ldots + c_r(a_{r1}v_1 + a_{r2}v_2 + \ldots + a_{rn}v_n) = 0$$

i.e., $(c_1a_{11} + c_2a_{21} + \ldots + c_ra_{r1})v_1 + (c_1a_{12} + c_2a_{22} + \ldots + c_ra_{r2})v_2$

$$+ \ldots + (c_1a_{1n} + c_2a_{2n} + \ldots + c_ra_{rn})v_n = 0$$

Since $v_1, v_2 \ldots, v_n$ are linearly independent, from (3) we have

$$a_{11}c_1 + a_{21}c_2 + \ldots + a_{r1}c_r = 0$$

$$a_{12}c_1 + a_{22}c_2 + \ldots + a_{r2}c_r = 0$$

$$\vdots \qquad \vdots \qquad \ldots \qquad \vdots$$

$$a_{1n}c_1 + a_{2n}c_2 + \ldots + a_{rn}c - r = 0$$

which is a homogenous system of n equations in r unknowns $c_1, c_2, \ldots, c_r$. Since $n < r$, this system has a nontrivial solution. Thus, at least one of the $c_i s$ is not zero and so the set B is linearly dependent.

Theorem. Let $v_1, v_2, \ldots, v_n$ be linearly independent in a vector space V over a field F. if v is any nonzero vector in V, then the set $\{v_1, v_2, \ldots, v_n, \text{v}\}$ is a linearly independent if and only if v is not in the linear span $\langle v_1, v_2, \ldots, v_n \rangle$.

Proof. Suppose that v is not in the linear span $\langle v_1, v_2, \ldots, v_n \rangle$ Consider any equation of the form

$$a_1v_1 + a_2v_2 + \ldots + a_nv_n + av = 0.$$

where $a_1, a_2, \ldots .a_n .a$ are scalars. If $a \neq 0$, then

$$v = -\frac{1}{a}(a_1v_1 + a_2v_2 + \ldots + a_nv_n)$$

Showing that $v \in \langle v_1, v_2, \ldots, v_n \rangle,$ contrary to the supposition unless $a = 0.$ But then, since $v_1, v_2, \ldots, v_n$ are linearly independent, equation (1), with $a = 0$, implies $a_i = 0,\ 1 \le i \le n.$ Thus $v_1, v_2, \ldots, v_n, v$ be linearly independent but $v \in \langle v_1, v_2, \ldots, v_n \rangle,$

Then, for some scalars $a_1, a_2, \ldots a_n, v$ can be expressed as a linear combination of $v_1, v_2, \ldots, v_n$

i.e., $v = a_1 v_1 + a_2 v_2 + \ldots + a_n v_n$

So, $a_1 v_1 + a_2 v_2 + \ldots + a_n v_n + (-1) v = 0$

That is, $v, v_1, v_2, \ldots, v_n$ are linearly dependent contrary to our assumption.

Hence $v \in \langle v_1, v_2, \ldots, v_n \rangle,$

Let V be an n dimensional vector spacr over a field $F.$ Then every set $S = \{v_1, v_2, \ldots, v_n\}$ *of* n linearly independent vectors in V is a basis for $V.$

Proof. Each of the vectors $v_1, v_2, \ldots, v_n$ is a linear combination of these n vectors, since every vector is equal to itself. *Let* $v \in V$ be any vector. The set of vectors $\{v_1, v_2, \ldots, v_n, v\}$ is linearly independent, by theorem Therefore, $a_1 v_1 + a_2 v_2 + \ldots + a_n v_n + av = 0$ implies at least on of the scalars $a_1, a_2, \ldots, a_n, a$ belonging to F is not zero. But $a \ne 0,$ since $v_1, v_2, \ldots, v_n$ are linearly independent. Therefore, v can be expressed as a linear combination of the vectors $v_1, v_2, \ldots, v_n$. Thus $V = \langle S \rangle$ and so S is a basis for $V.$

Theorem.

(i) Any linearly Independent set f vectors in a finite dimensional vector space V can be extended to a basis for $V.$

(ii) if W is a subspace of a finite dimension vectors space V, then dim $W \leq \dim V$ *Moreover*, if dim $W = \dim V$, *then* $W = V$.

Proof.

(i) $Let\, S = \{v_1, v_2, \ldots, v_r\}$ be a linearly independent set of vectors in an $n-dimensional$ vectors space V and $r < n$. Since dim $V = n$, the set S cannot span V There is a vectors, sy $v_{r+1} \in V$, such that $v_{r+1} \notin \langle S \rangle$. By theorem the set $\{v_{r+1}\} \cup S$ is linearly independent. The process can be repeated $n-r$ times to get a larger set $\{v_1, v_2 \ldots, v_{r+1}, \ldots, v_n\}$ which is linearly independent. By Theorem, this set is a basis for V.

(ii) Let V be of dimension n. Then, by any set of $n+1$ or more vectors is linearly dependent. Furthermore, sine a basis of W consist of linear independent vectors, it cannot contain more than n elements. Hence dim $W \leq n = \dim V$.

If dim $W = \dim V$, *then every basis* W is also a basis for V. Thus $W = V$.

The following theorem gives an equivalent definition of direct sum

Theorem. A vector space V is the direct sum of its subspace $U\, and\, W$ if and only if each $v \in V$ can be written uniquely as (i) $v = u + w, \quad u \in U, \quad w \in W$.

Proof. Recall that a vector space V is the direct sum of its subspace $U\, and\, W$ if

$(\mathbf{i})\, \mathrm{V} = \mathrm{U} + \mathrm{W}$ and $(\mathbf{ii})\, \mathrm{U} \cup \mathrm{W} = \{0\}$.

Suppose that V is the direct sum of its subspace $U\, and\, W$. Then conditions (i) and (ii) satisfied.

Now let $v \in V$ Then

$v = u + w$

For some $u \in U, w \in W$, To see that the expression for v in (1) is unique, let

$v = u_1 = w_1 - w \in U \cap W = \{0\}$.

Hence $u - u_1 = w - w_1$. That is $u = u_1, w = w_1$. So the expression unique and (i) is satisfied

Conversely, suppose that (i) is satisfied. Then (i) is part of (i). To see that (ii) is satisfied, let $v \in U \cap W .Then v$ can be written as

$v = u + 0 \quad ,u \in U$

$= 0 + w \quad ,w \in W$

By (i)', the expression for v is unique. Therefore

$u = 0 = w, so that\, v = 0.$

Hence $U \cap W = \{0\}$

Theorem. If $U\, and\, W$ are finite dimensional subspace of a vector space V over a field, F, then

(i) $\dim(U + W) = deim U + \dim W - \dim(U \cap W)$.

(ii) if $U \cap W = \{0\}\ then V = U \oplus W$ and $\dim V = \dim U + \dim W$.

Proof. (i) Suppose $U \cap W \neq \{0\} and\, U \cap W$ has a basis

$\{v_1, v_2, \ldots, v_r\}\, Since U \cap W \subset U$, this set of vector from a part of a basis of W. Thus $U \cap W, U, W$ have dimension $r, r+s, r+t$ respectively.

Clearly $\{v_1, v_2, \ldots, v_r, u_1, u_2, \ldots, u_s, w_1, w_2, \ldots, w_t\}$. This will follow if we show that this set is linearly independent and hence forms a basis for $U + W$.

Next suppose that $U \cap W = \{0\}.Let\ \{u_1, u_2, \ldots, u_s\}$ be basis for U and $\{w_1, w_2, \ldots, w_t\}$ be a basis for W. Then $\{u_1, u_2, \ldots, w_1, w_2, \ldots, w_t\}$ is a spanning set for $U + W = U \oplus W$.

Now if $a_1u_1+\ldots+a_su_s+b_1w_1+\ldots+b_sw_s=0$,

Then $a_1u_1+a_2u_2+\ldots+a_su_s=-(b_1w_1+b_2w_2+\ldots+b_tw_t)=0$ since each vector is I $U\cap W$. This in turn implies that $a_1=\ldots=a_s=b_1=\ldots=b_t=0$. Hence $\{u_1,u_2,\ldots,u_s,w_1,w_2,\ldots,w_t\}$ is a basis for $U\oplus W$ and $\dim(U\oplus W)=s+t=\dim U+\dim W$.

This agree with the general result with $\dim(U\cap W)=0$

Here we define the dimension of the vector space $\{0\}$ *to be zero.*

ROW AND COLUMN SPACES

Definition. Let $A=[a_{ij}]$ be an $m\,xn$ matrix over $R, m\le n.$ As defined in row of A $P_1=[a_{11}\,a_{12}\ldots a_{1n}],\ p_2=[a_{21}\,a_{22}\ldots a_{2n}],\ldots,$ $P_m=[am_1\ am_2\ldots a_{mn}]$ are called row vectors of A. The subspace $\boldsymbol{R}^n$ spanned by $\{p_1,p_2,\ldots,p_m\}$ is called the row space of the matrix A. Similarly, the subspace spanned by the column vectors

$$\begin{bmatrix} a_{11} \\ a_{21} \\ \vdots \\ a_{m1} \end{bmatrix}, \begin{bmatrix} a_{12} \\ a_{22} \\ \vdots \\ a_{mn} \end{bmatrix}, \ldots, \begin{bmatrix} a_{1n} \\ a_{2n} \\ \vdots \\ a_{mn} \end{bmatrix}$$

Of A is called the column space of A. The dimension of the row space is called the row rank of the matrix A. Thus the row rank is the maximum number of linearly independent row vectors of A, similarly ,the column rank of A is the maximum number of linearly independent column vector of A.

Theorem. Row equivalent matrices have the same row space.

Proof. $Let\, B\, \tilde{R}\, A.$ Then, since rows of B are obtained from the rows of A by elementary roe operations each row of B is clearly a row of A or a linear combination of rows of $A.$ Hence the row space of B is contained in the row space of A .On the other hand, we can apply inverse elementary row ***operations***[1] to B and obtain A. Hence the row space of A is contained in the row space of B. Thus, row space of $A = row$ space B.

Corollary. Row equivalent matrices have the same row rank.

Corollary. If P is nonsingular, then the matrices A and PA have the same row rank. As $PA\tilde{R}A,$ the row rank of *PA* is equal to the row rank of A.

Corollary. The row rank of a nonsingular matrix A of order $n\, x\, n\, is\, n.$

Proof. Since A is nonsingular, $A\,\tilde{R}\,I_n.$ The n rows of I_n are linearly independent, because

$$a_1[1\ 0\ 0\ldots0]+a^2[0\ 100\ldots0]+\ldots+a_n[00\ldots01]=[0\ 0\ldots\ 0]$$

$$a_1 = 0,\ \ a_2 = 0,\ldots,a_n = 0$$

Hence row rank of I_n is n and so row rank of A is n.

In this we continuous our study of the structure of a vector space V by determination a smallest set of vectors in V that completely describe V.

Definition. The vector $v_1, v_2, \ldots, v_n$ in a vector space V are said to from a basis for V if $(a)\, v_1, v_2, \ldots v_k$ are linearly independent.

Remark. if $v_1, v_2, \ldots v_k$ form a basis for vector space $V,$ then they must be distinct and nonzero , so we write them as a set $\{v_1, v_2, \ldots v_k\}.$

Example. The vector $e_1 = (1,0)\, and\, e_2 = (0,1)$ from a basis for R^2 the vectors e_1 ,$e_2\, and\, e_3$ from basis for R^3 and in general , the vectors $e_1, e_2, \ldots e_n$ form a basis for $R^n.Each$ of these sets vectors is called the

natural basis or **standard basis** for $R^2.R^3.and\, R^n$. Respectively.

Example. Show that the set

$S=\{v_1,v_2,v_3,v_4\}.where\, v_1=(1,0,1,0).v_2=(0,1,-1,2)$

. $v_3=(0.2.2.1).and\, v_4=(1,0,0,1)$ is a basis for R^4.

Solution. To show that S is linearly independent, we form the equation

$c_1v_1+c_2v_2+c_3v_3+c_4v_4=0$

And solve for $c_1,c_2,c_3\, and\, c_4$. Substituting for $v_1,v_2,v_3\, and\, v_4$. we obtain the linear system (verify)

$c_1\quad +c_4=0$

$c_2+2c_3=0$

$c_1-c_2+2c_3=0$

$2c_2+c_3+c_4=0.$

which has a its only solution $c_1=c_2=c_3=c_4=0(verify).$ Showing that S is linearly independent .

To show that S spans R^4, we let $v=(a,b,c,d)\, be\, any\, vector\, in\, R^4$. We now seek constant $k_1,k_2,k_3\, and\, k_4\, such\, that$

$k_1v_1+k_2v_2+k_3v_3+k_4v_4=v.$

Substituting for $v_1,v_2,v_3,v_4\, and\, v,$ we find a solution (verify) for $k_1,k_2,k_3,$ and k_4 to the resulting linear system for any a,b,c and d. Hence S span R^4 and is a basis for R^4.

Example. Show that the set $S=\{t^2+1.t-1,2t+2\}$ is a basis for the vector space P_2.

Solution. To do this, we must that S spans V and is linearly independent. To show that it spans. $V,$ we take any vector $in\, V,$ that is, a polynomial $at^2+bt+c.$ and must find constant $a_1, a_2 .and\ a_3$ such that

$at^2+bt+c = a_1(t^2+1)+a_2(t-1)+a_3(2t+2)$

$= a_1t^2+(u_2+2a_3)t+(a_1-a_2+2a_3).$

Since two polynomials agree for all values of t only if the coefficient or respective power of t agree, we get the linear system

$a_1 = a$

$a_2+2a_3 = b$

$a_1-a_2+2a_3 = c.$

Solving, we have

$$a_1 = a,\ a_2 = \frac{a+b-c}{2},\ a_3 = \frac{c+b-a}{4}$$

Hence S spans V.

To illustrate this result, suppose that we are given the vector $2t^2+6t+13.$ Here $a=2, b=6\,and\,c=13.$ Substituting in the foregoing expressions for $a_1, a_2\ and\ a_3.$ we find that

$$a_1 = 2.\ a_2 = -\frac{5}{2},\ a_3 = \frac{17}{4}.$$

Hence $2t^2+6t+13 = 2(t^2+1)-\frac{5}{2}(t-1)+\frac{17}{4}(2t+2)$

To show that S is linearly independent, we form $a_1(t^2+1)+a_2(t-1)+a_3(2t+2)=0.$

Then, $a_1t^2+(a_2+2a_3)t+(a_1-a_2+2a_3)=0.$

Again, this can hold for all values of $t\,only$ if

$a_1 \qquad = 0$

$a_2+2a_3 = 0$

$a_1-a_2+2a_3 = 0.$

The only solution to this homogenous system is $a_1 = a_2 = a_3 = 0,$ which implies that S is linearly independent. Thus $S\, is\, a\, basis$ for P_2.

The set of vectors $\{t^n . t^{n-1}, \ldots .t, 1\}$ forms a basis for the vector space p_n called the natural. Or standard basis, for P_n. it has already been shown in that this set of vectors span P_n. Linear independence is left as an exercise.

Example. Find a basis for the subspace V of P_2, consisting of all vectors of the form $at^2 + bt + c, where\, c = a - b.$

Solution. Every vector in V is of the form $at^2 + bt + a - b$

which can be written as $a(t^2 + 1) + b(t - 1)$

So the vector $t^2 + 1 and\; t - 1 span V$. Moreover, these vectors are linearly independent because neither one is a multiple of the other. This conclusion could also be reached (with more work) by writing the equation.

$a_1(t^2 + 1) + a_2(t - 1) = 0$ or

$t^2 a_1 + t a_2 + (a_1 - a_2) = 0.$

Since the equation is to hold for all values of t. we must have $a_1 = 0$ and $a_2 = 0.$

A vector space V is called finite-dimensional if there is a finite-dimensional. If there is a finite subset of V that is a basis for V. if there is no such finite subset of $V, then V$ is called infinite-dimensional.

Almost all the vector spaces considered in this book are finite dimensional. However, we point out that there are many infinite – dimensional vector spaces That there are many infinite-dimensional vector spaces that are extremely important in mathematics and physics. Their study lies beyond the scope of this book. The vector space $P,$

consisting of all polynomials, and the vector space $C(-\infty,\infty)$ is defined of section are not finite-dimensional.

We now establish some results about finite-dimensional vector spaces that will tell about the number of vectors in a basis, compare two different basis, and gives properties of bases. First, we observe that if $(cv_1, v_2 \ldots, v_k\}$ is also a basis when $c \neq 0$. Thus a basis for a nonzero vector space is never unique.

Theorem. if $S = \{v_1, v_2, \ldots v_n\} is\, a\, basis$ for a vector space $V\cdot$ then ever in V can be written in one and only one way as a linear combination of the vectors in S.

Proof. First, every vector **v** in V can be written as a linear combination of the vectors in S because S spans. $V\cdot$ Now left

$$v = c_1 v_1 + c_2 v_2 + \ldots + c_n v_n$$

and

$$v = d_1 v_1 + d_2 V_2 + \ldots + d_n v_n.$$

Subtracting, we obtain

$$0 = (c_1 - d_1) v_1 + (c_2 - d_2) v_2 + \ldots + (c_n - d_n) v_n.$$

Since S is linearly independent , it follows that $c_i - d_i = 0, 1 \leq i \leq n,$ so $c_i = d_i, 1 \leq i \leq n.$ Hence there is only one way to express v as a linear combination of the vectors in $S\cdot$ **Proof.** First, every vector $v in V\cdot$ can be written as a linear combination of the vectors in $S\, because\, S$ spans. V Now let

$$v = c_1 v_1 + c_2 v_2 + \ldots + c_n v_n$$

and

$$v = d_1 v_1 + d_2 v_2 + \ldots + d_n v_n.$$

Subtracting from , we obtain

$$= (c_1 - d_1) v_1 + (c_2 - d_2) v_2 + \ldots + (c_n - d_n) v_n.$$

Since S is linearly independent, it follows that $c_i = d_i, 1 \le i \le n.$ Hence there is only one way to express v as a linear combination of the vectors in S.

We can also show that if $S = \{v_1, v_2, \ldots, v_n\}$ *is a set* of nonzero vectors in a vector space V such that every vector in V can be written in one and only one way as a linear combination of the vectors in S, then S is a basis for V.

Theorem. Let $S = \{v_1, v_2, \ldots, v_n\}$ is a set of a nonzero vectors in a vector space V and let W=span S Then some subset of S is a basis for W.

Proof. Case 1. If S is linearly independent then Since S already spans W, we conclude than S is basis for

Case II. If S is linearly dependent, then

$$c_1 v_1 + c_2 v_2 + \cdots + c_n v_n = 0.$$

where c_1, c_2, ...c_n are not all zero. Thus, some v_j from getting a subset S_1 of S. Then by the observation made just before in We conclude that $S_1 = \{v_1, v_2, \ldots v_{j-1}, v_{j+1}, \ldots \ldots . v_n\}$ also spans W . The Set T is a basis for W.

If S_1 is linearly independent, then S_1 is a basis if S_1 is linearly dependent, delete a vector of s_1 that is a linear combination of the preceding vectors of S_1 and get a new set S_2 which spans W Continuing, since S is a finite set , we will eventually find a subset T of S that is linearly independent and spans W The set T is a basis f for W.

Alternative Constructive proof when V is R^m, n≥m We take the vectors in S as m×1 matrices and from above. This equation leads to a homogenous system in the n unknowns c_1, c_2, ...c_n. The columns of its m×n coefficient matrix $A\ are\ v_1, v_2, \ldots v_n$.We now transform A to a matrix B in reduced row echelon form. Having r leading 1`in the r nonzero rows, of B occur in the first r columns. Thus, we have

$$B = \begin{bmatrix} 1 & 0 & 0 & \dots & 0 & b_{1r+1} & \dots & b_{1n} \\ 0 & 1 & 0 & \dots & 0 & b_{2r+1} & \dots & b_{2n} \\ 0 & 0 & 1 & \dots & 0 & b_{3r+1} & \dots & b_{3n} \\ \vdots & \vdots & \vdots & \ddots & \vdots & \vdots & & \vdots \\ 0 & 0 & 0 & \dots & 1 & b_{rr+1} & \dots & b_{rn} \\ 0 & 0 & 0 & \dots & 0 & 0 & \dots & 0 \\ \vdots & \vdots & \vdots & \ddots & \vdots & \vdots & & \vdots \\ 0 & 0 & 0 & \dots & 0 & 0 & \dots & 0 \end{bmatrix}$$

Solving for the unknowns corresponding to the leading $1's,$ We see that $c_1, c_2, \dots, c_r$ can be solved for in terms of the other unknowns $c_{r+1}.c_{r+2} \dots .c_n.$ Thus

$$c_1 = -b_{1_{r+1}} c_{r+1} - b_{1_{r+2}} c_{r+2} - \dots - b_{1_n} c_n$$

$$c_2 = -b_{2_{r+1}} c_{r+1} - b_{2_{r+2}} c_{r+2} - \dots - b_{2_n} c_n$$

$$\vdots$$

$$c_r = -b_{r_{r+1}} c_{r+1} - b_{r_{r+2}} c_{r+2} - \dots - b_{r_n} c_n,$$

Where $c_{r+1}+, c_{r+2}, \dots, c_n$ can be assigned arbitrary real values. Letting

$$c_{r+1} = 1. \quad c_{r+2} = 0 \dots \dots . c_n = 0.$$

In equation and using these values in equation we have

$$-b_{1r+1}v_1 - b_{2r+1}v_2 - \dots - b_{r\,r+1}\, v_r + v_r + v_{r+1} = 0$$

which implies that v_{r+1} is a linear combination of $v_1, v_2, \dots, v_r.$ By the observation made just before the set of vectors obtained form S by deleting v_{r+1} spans $W.$ Similarly , letting

$$c_{r+1} = 0. \quad c_{r+2} = 1. c_{r+3} = 0. \dots\dots c_n = 0$$

We find that v_{r+2} is a linear combination of $v_1, v_2 \dots\dots v_r$ and the set of vectors obtained from S by deleting v_{r+1} spans $W.$ continuing in this manner, $v_{r+3}, v_{r+4}, \dots\dots v_n$ are linear combinations $v_1.v_2, \dots\dots v_r,$ so it follows that $\{v_1.v_2,......v_r,\}$ spans W.

We next show that $\{v_1.v_2,......v_r,\}$ is linearly independent. Consider the matrix B_D obtained by deleting from B all columns not containing a leading 1. In this case. B_D consists of the first r columns of B. Thus,

$$B_D = \begin{bmatrix} 1 & 0 & 0 \ldots & 0 \\ 1 & 0 & 0 \ldots & 0 \\ 0 & 0 & 1 & \\ 0 & 0 & \ldots & 1 \\ 0 & 0 & \ldots & 0 \\ \vdots & \vdots & & \vdots \\ 0 & 0 & \ldots & 0 \end{bmatrix}$$

Let A_D be the matrix obtained from A by deleting the columns corresponding to the columns that were deleted in B to obtain B_D. In this case, the columns of A_D are $v_1.v_2,......v_r, the\ first\ r$ columns of A. Since $A\ and\ B$ are row equivalent. So are $A_D\ and\ B_D$. Then the homogeneous systems

$$A_D x = 0 \quad and \quad B_D x = 0$$

are equivalent. Recall now that the homogenous system $B_D x = 0$ can be written equivalently as $x_1 w_1 + x_2 w_2 + \ldots + x_r\ w_r = 0$
where

$$x = \begin{bmatrix} x_1 \\ x_2 \\ \vdots \\ x_r \end{bmatrix}$$

and $w_1, w_2 ......w_r$ are the columns of B_D. Since the columns of B_D form a linearly independent set of vector in R^m Equation has only the trivial solution. Hence $A_D x = 0$ also has only the trivial solution. Thus, the columns of A_D are linearly independent. That is, $\{v_1.v_2,......v_r,\}$ is linearly independent. The first proof of Theorem leads to a simple procedure for finding a subset T of $a\ set\ S\ so\ that\ T\ is\ a\ basis\ for\ span\ S$.

Let $S = \{v_1.v_2,......v_n,\}$ be a set of nonzero vectors in a vector space V The procedure for finding subset of S that is a basis for $W = span\,S$ is as follows.

Step 1. From Equation $c_1 v_1 + c_2 v_2 + \ldots + c_n v_n = 0.$

We solve for $c_1\ , c_2 ......c_n.$ if these are all zero, then S is linearly independent an is than a basis for $W.$

Step 2. If $c_1\ , c_2 ......c_n.$ if these are all zero, then S is linearly independent, so one of the vectors in S we obtain a subset $T\ and\ S$ that spans W and is linearly independence. Thus T is a basis for $W.$

This procedure can rather be tedious, since every time we delete a vector from $S,$ we must solve a linear system. We shall present a much more efficient procedure for finding a basis for $W = spans$, but the basis is not guaranteed to be a subset of $S.$ In many cases this is not a cause for concern, since one basis for $W = span\,S$ is good as any other basis .However. there are cases when the vectors in S have some special properties and we want the basis for $W = span\,S$ to have the same properties, so we want the basis to be a subset of $S.$ if $V = R^m$, the alternative proof of Theorem Yields a very efficient procedure for finding a basis for $W = span\,S$ consisiting of vector form $S.$

And $w_1, w_2 ......w_r$ are the columns of $B_D.$ Since the columns of B_D form a linearly independent set of vector in R^m Equation has only the trivial solution. Hence $A_D x = 0$ also has only the trivial solution. Thus the columns of A_D are linearly independent. That is, $\{v_1.v_2,......v_r,\}$

is linearly independent. The first proof of Theorem leads to a simple procedure for finding a subset T of $a\, set\, S\, so\, that\, T\, is\, a\, basis\ \ for\, span\, S.$

Let $S=\{v_1.v_2,\ldots\ldots v_n,\}$ be a set of nonzero vectors in a vector space V The procedure for finding subset of S that is a basis for $W=span\,S$ is as follows.

Let $V=R^m$ and let $S=\{v_1,v_2,\ldots,v_n]$ be a set of nonzero vector in V The procedure for finding a subset of S that is a basis for $W=span\,S$ is as follows.

Step 1. From equation, $c_1v_1+c_2v_2+\ldots+c_nv_n=0.$

Step 2. Construct the augmented matrix associated with the homogeneous system of equation and is transform it to reduce row echelon form.

Step3. The vector corresponding to the columns containing the leading $1's$ form a basis for $W=span\,S.$ Recall that in the alternative proof of we assumed without loss of generally that the r leading $1's$ occur in columns 1,3 and 4, then $\{v_1,v_3,v_4\}$ is a basis for span $S.$

Remark In step 2 of the procedure above, it is sufficient to transform the augmented matrix to row echelon form

Example. Let $S=\{v_1,v_2,v_3,v_4,v_5\}$ be a set of vector R^4 , where

$v_1=(1,2-2,1)\quad v_2=(-3,0,-4,3),$

$v_3=(2,1,1-1),\ v_4=(-3\ ,3,-9,6)$ and $v_5=(9,3,7,-6).$

Find a subset of S that is a basis for $W=span\,S.$

Solution

Step 1 form equation

$c_r(1,2,-2)+c_2(-3,0\ ,-4,3)+c_3(2,1,1,1-1)+c_4(-3,3,-9\ ,6)$

$c_5(9,3,7-6)=(0,0,0,0),$

Step 2. Equating corresponding component, we obtain homogenous system

$c_1 - 3c_2 + 2c_3 - 3c_4 + 9c_5 = 0$

$2c_1 \quad + c_3 + 3c_4 + 3c_5 = 0$

$-2c_1 - 4c_2 + c_3 - 9c_4 + 7c_5 = 0$

$c_1 = 3c_2 - c_3 + 6c_4 - 6c_5 = 0.$

The reduced row echelon form of the associated matrix is verify

$$\begin{bmatrix} 1 & 0 & \frac{1}{2} & \frac{3}{2} & \frac{3}{2} & \vdots 0 \\ 0 & 1 & -\frac{1}{2} & \frac{3}{2} & -\frac{5}{2} & \vdots 0 \\ 0 & 0 & 0 & 0 & 0 & \vdots 0 \\ 0 & 0 & 0 & 0 & 0 & \vdots 0 \end{bmatrix}$$

Step 3. The leading 1's appear in columns 1 and 2, so $\{v_1, v_2\}$ is a basis for $W = span\, S.$

Remark. In the alternative proof of theorem when $v = R^n$, the order of the vectors in the original spanning set S determines which for W is obtained. If for example, we consider when $S = \{\mathbf{w}_1, \mathbf{w}_2, \mathbf{w}_3, \mathbf{w}_4, \mathbf{w}_5\}$ with $\mathbf{w}_1 = \mathbf{v}_4,\ \mathbf{w}_2 = \mathbf{v}_3 . \mathbf{w}_3 = \mathbf{v}_2, \mathbf{w}_4 = \mathbf{v}_1, and\ \mathbf{w}_5 = \mathbf{v}_5$

$$\begin{bmatrix} 1 & 0 & \frac{1}{3} & \frac{1}{3} & -\frac{1}{3} & \vdots 0 \\ 0 & 1 & -1 & 1 & 4 & \vdots 0 \\ 0 & 0 & 0 & 0 & 0 & \vdots 0 \\ 0 & 0 & 0 & 0 & 0 & \vdots 0 \end{bmatrix}$$

it now follows that $\{w_1, w_2\} = \{v_4, v_3\}\, is\, a\, basis\ for\ w = span\, S.$

We shall soon establish a main result of this section, which will tell us about the number of vectors in two different bases. First observe that if $\{v_1, v_2 \ldots\ldots v_n\}$ is a basis for a vector space $V \cdot$ then $\{cv_1, v_2, \ldots\ldots v_n\}$ is also a basis if $c \neq 0$. Thus a vector space always has infelicity many bases.

Theorem. if $S=\{v_1,v_2\ldots\ldots v_n\}$ is a basis for a vector space $V\ and\ T=\{w_1,w_2\ ,\ldots\ldots,w_r\}$ is a linearly independent set of vectors in $V,$ then r $\leq n.$

Proof. Let $t_1=\{w_1,v_1,\ldots\ldots,v_n\}$. Since S spans $V,$ so does $T_1.$ Since w_1 is a linear combination of the vector in S, we find that T_1 is linearly dependent. Then, by theorem some v_j is linear combination of the preceding vectors in T_1 Delete that particular vector $v_j.$

Let $S_1=\{w_1.v_1\ ,\ldots\ldots v_{j-1},v_{j+1}\ldots\ldots v_n$. Note that S_1 spans $V.$ Next let $T_2=\{w_2,w_1,v_1\ldots\ldots v_{j-1},v_{j+1}\ ,\ldots\ldots v_n\}$. Then T_2 is linearly dependent and some vector in T_2 is a linear combination of the preceding vectors in T_2. Is a linear independent this vector cannot be w_1, so it is $v_i\, i\neq j.$ Repeat this process over and over . if the v vectors are all eliminated before we can run out of w vectors, then the resulting set of w vectors, a subset of T, is linearly dependent , which implies that T is also linearly dependent. Since we have reached a contradiction, we conclude that the number r of w vectors must be no greater than the number $n\, of\, v$ vectors. That is, $r\leq n.$

Corollary. If $S=\{v_1,v_2,\ldots\ldots v_n\}\, and\, T-=\{w_1,w_2,\ldots\ldots w_m\}$ are basis for a vector space, than $n=m.$

Proof. Since T is linearly independent set of vectors, implies that $m\leq n.$ Similarly $n\leq m\, because\, S$ is linearly independent. Hence $n=m.$

Thus, although a vector space has many bases we have just shoen that for a particular vector space $V,$ all bases have the same number of vectors. We can then the following definition.

Dimension: - The dimension of a nonzero vector space V is the number of the vectors in a basis for V. we often write **dim** V for the dimension of $V.$ since set $\{0\}$ is linearly depedent, it is natural to say that the vector space $\{0\}$ has dimension zero.

Example. The dimension of R^2 is 2; the dimension of R^3 is 3; and in general, the dimension of R^n is n

Example. The dimension of P_2 is 3; the dimensional of P_3 is 4; and in general, the dimension of $P_n\, is\, n+1$. It can be shown that all finite-dimension vector spaces of the same dimension differ only in the nature of their elements; their algebraic properties are identical.

It can also be shown that if V is a finite dimensional vector space then every nonzero subspace $W\, of\, V$ has a finite basis and dim $W \leq \dim V$

Example. The subspace $W\, of\, R^4$ n considered in has dimension 2.

We might the subspace of R^2 [recall that R^2 can be viewed as the (x, y) - plane]. First, we have $\{0\}$ and R^2, the trivial subspace of dimension 0 and 2 respectively, Now the subspace V of R^2 spanned by a vector $v \neq 0$ is one –dimensional subspace of R^2; V is represented by line through the orign. Thus the subspace of R^2 are $\{0\}, R^2$,and all the line through the origin. Similarly, we ask you to show that the subspace of $R^3\, are \{0\}.R^3$, all lines through the origin, and all planes through the origin.

It can be shown that if a vector space V has dimension n. Then any set of $n+1$ vector $in V$ is necessarily linearly dependent. Any set of more than n vectors in R^n is linearly dependent. Thus the four vectors in R^3 considered in were shown to be linearly dependent. Also, if a vector space V can span V were shown linearly dependent It can be shown that if a vector space V has dimension n. Then any set of $n+1$ vector $in V$ is necessarily linearly dependent. Any set of more than n vectors in R^n is linearly dependent. Thus the four vectors in R^3 considered in were shown to be linearly dependent. Also, if a vector space V can span V were shown linearly dependent.

We now come to a Theorem that we shall have occasion to use several times in constructing a basis containing a given set of linearly independent vectors. We shall leave the proof as an exercise. The example following the theorem completely unities the proof.

Theorem. If S is linearly independent set of vectors in a finite-dimensional vector space V, then there is a basis T for V, which contain S.

Theorem says that a linearly independent set of vector in a vector space V can be extended to a basis for V to a basis for V, which contain S.

Example. Suppose that we wish to find a basis for R^4 that contain the vectors $v_1 = (1\ ,0,1,0)\, and\, v_2 = (-1,1,-1,0)$.

We use First let $\{e_1, e_2, e_3, e_4\}$ be the natural basis for R^4, where

$e_1 = (1,0,0,0),\ e_2 = (0,1,0,0),\ e_3 = (0,0,1,0).$

and

$e_4 = (0,0,01).$

From the set $S = \{v_1, v_2, e_1, e_2, e_3, e_4\}\, spans\, R^4$, so does S. we now use the alternative proof of to find a subset of S that is a basis for R^4. Thus, we form Equation,

$c_1v_1 + c_2v_2 + c_3e_1 + c_5e_3 + c_6e_4 = 0,$

Which leads to the homogenous system

$c_1 - c_2 + c_3 = 0$

$-c_2 \quad c_4 = 0$

$c_1 - c_2 \quad +c_5 = 0$

$c_6 = 0.$

Transforming the augmented matrix to reduce echelon form, we obtain (verify)

$$\begin{bmatrix} 1 & 0 & 0 & 1 & 1 & 0 & \vdots 0 \\ 0 & 1 & 0 & 1 & 0 & 0 & \vdots 0 \\ 0 & 0 & 1 & 0 & -1 & 0 & \vdots 0 \\ 0 & 0 & 0 & 0 & 0 & 1 & \vdots 0 \end{bmatrix}$$

Since the leading $1's$ appear in columns $1,2,3$ and 6, we conclude that $\{v_1, v_2, e_1.e_4\}$ is a basis for R^4 containing $v_1 \; and \; v_2$

From the definition of a basis, a set of vectors in a vectors in a vector space V is basis for V if it spans V and is linearly independent. However, if we are given the additional information that the dimension of V is $n,$ we need only verify one of the two conditions. This is the content of the following theorem.

Theorem.

Let V be an $n-$ dimensional vector space, and $Let\, S = \{v_1, v_2, \ldots\ldots, v_n\}$ be a set of a vectors in $V\cdot$

(a) if S is linearly independent, then it is a basis for $V\cdot$

(b) $if\, S$ spans V ,then it is a basis for $V\cdot$

Proof. Exercise

As in a particular application of theorem, we have the following. To determine if a subset $S\, of\, R^n$ first count the number of elements n $S\cdot$ if S has n elements, we can use either part (a) or part (b) of to determine whether S is or is not a basis. $if\, S$ does not have elements, it is not a basis for R^n . The same line of reasoning applies to any vector space or subspace whose dimension is known.

Example. $W = span\, S$ is a subspace of R^4, sp dim $W \leq 4\cdot$ Since S contain five vectors, we conclude by Corrollary that S is not a basis for $W\cdot$ since dim $R^4 = 4$ and the set S contains four vectors. It is possible for S to be a basis for R^4. $If\, S$ is linearly independent or spans. $R^4,$ It is a basis; otherwise it is not a basis. Thus, we need only check one of the condition in not both.

RANK-ALTERNATIVE METHOD

The rank of a matrix A was found by reducing it to on echelon form. To reduce A to an echelon from many operations are needed and in the process fractions creep in which make the computations awkward and cumbersome.

The Alternative Method is very elegant involving no fractions and it also yields an echelon matrix that is row equivalent to and whose nonzero rows constitute a basis for the row space of A..

Definition. Let $A=\left[a_{ij}\right]$ be an $m\times n$ matrix. We define determination d_{ij} of order 2 as under:

For $2\le i\le m\, and\, 2\le j\le n,$

$$d_{ij}=\det\begin{bmatrix} a_{11}\, a_{ij} \\ a_{i1} \quad a_{ij} \end{bmatrix}=\begin{bmatrix} a_{11}\, a_{ij} \\ a_{i1} \quad a_{ij} \end{bmatrix}=a_{11}a_{ij}-a_{ij}a_{i1}$$

Theorem. Let $A=\left[a_{ij}\right] be\, an\, m\times n$ matrix with $a_{11}\ne 0\, and\, d_{ij}$ be as in Then

$$Rank\ A=1+Rank\begin{bmatrix} d_{22} & d_{23}\ldots d_{2n} \\ d_{32} & d_{33}\ldots\ d_{3n} \\ \vdots & \vdots\ \ \ldots\ \ \ \vdots \\ d_{m2} & d_{m3}\ldots d_{mn} \end{bmatrix}$$

Proof. Let

$$A=\begin{bmatrix} a_{11} & a_{12} & \cdots & a_{1n} \\ a_{21} & a_{22} & \cdots & a_{2n} \\ \vdots & \vdots & \ldots & \vdots \\ a_{m1} & am_{1} & \cdots & a_{mn} \end{bmatrix} be\, an\ m\, x\, n\ matrix\, and\, a_{11}\ne 0.$$

Multiply $R_i,\left(2\le i\le m\right) by\, a_{11}\, so\, that$

$$A\tilde{R}\begin{bmatrix} a_{11} & a_{12} & \cdots & a_{1n} \\ a_{11}a_{21} & a_{11}a_{22} & \cdots & a_{11}a_{2n} \\ \vdots & \vdots & \cdots & \vdots \\ a_{11}am_1 & a_{11}am_2 & \cdots & a_{11}a_{mn} \end{bmatrix}$$

$$\tilde{R}\begin{bmatrix} a_{11} & a_{12} & \cdots & a_{1n} \\ 0 & a_{11}a_{22}-a_{12}a_{21} & \cdots & a_{11}a_{2n}-a_{1n}a_{21} \\ \vdots & \vdots & \cdots & \vdots \\ 0 & a_{11}a_{m2}-a_{12}a_{m1} & \cdots & a_{11}a_{mn}-a_{1n}a_{m1} \end{bmatrix} \text{ by } R_i - a_{i1}R_1 \ 2 \le i \le m$$

$$\tilde{R}\begin{bmatrix} a_{11} & a_{12} & \cdots & a_{1n} \\ 0 & d_{22} & \cdots & d_{2n} \\ 0 & d_{32} & \cdots & d_{3n} \\ \vdots & \vdots & \cdots & \vdots \\ 0 & d_{m2} & \cdots & d_{mn} \end{bmatrix}$$

Since $a_{11} \neq 0,$ the first row in the last matrix is not a linear combination of the other row and so the rank of

$$\begin{bmatrix} d_{22} & d_{23} & \cdots & d_{2n} \\ d_{32} & d_{33} & \cdots & d_{3n} \\ \vdots & \vdots & \cdots & \vdots \\ d_{m2} & d_{m3} & \cdots & d_{mn} \end{bmatrix}$$

As desired

The theorem will be used recursively until we obtain a matrix having one row (column)

An echelon matrix row equivalent to A will be obtained by retaining R_1 of A and replacing the rows $R_{i,}\ 2 \le i \le m,$ by the first b rows of the successive matrices obtained by recursive use of the theorem provided that the first $i-1$ entries in R_i are zero $(2 \le i \le m)$.

The method is illustrated by examples.

Example. Find the rank of the matrix

$$A = \begin{bmatrix} -3 & 5 & 1 & 2 \\ 7 & 2 & 0 & -4 \\ -8 & 3 & 1 & 6 \end{bmatrix}$$

Also write an echelon matrix row equivalent to

Solution.

$$A = 1 + Rank \begin{bmatrix} \begin{vmatrix} -3 & 5 \\ 7 & 2 \end{vmatrix} & \begin{vmatrix} -3 & 1 \\ 7 & 0 \end{vmatrix} & \begin{vmatrix} -3 & 2 \\ 7 & -4 \end{vmatrix} \\ \begin{vmatrix} -3 & 5 \\ -8 & 3 \end{vmatrix} & \begin{vmatrix} -3 & 1 \\ -8 & 1 \end{vmatrix} & \begin{vmatrix} -3 & 2 \\ -8 & 6 \end{vmatrix} \end{bmatrix}$$

Rank

$$= 1 + Rank \begin{bmatrix} -41 & -7 & -2 \\ 31 & 5 & -2 \end{bmatrix}$$

$$= 2 + Rank \begin{bmatrix} \begin{vmatrix} -41 & -7 \\ 31 & 5 \end{vmatrix} & \begin{vmatrix} -41 & -2 \\ 31 & -2 \end{vmatrix} \end{bmatrix}$$

An echelon matrix row equivalent to A is

$$\begin{bmatrix} -3 & 5 & 1 & 2 \\ 0 & -41 & -7 & -2 \\ 0 & 0 & 12 & 144 \end{bmatrix}$$

Example. Find the rank of the matrix

$$A = \begin{bmatrix} 1 & -2 & 1 & 0 & 5 \\ -1 & 0 & 1 & -2 & 2 \\ 1 & -6 & 3 & -2 & 12 \\ 2 & -2 & 0 & 2 & 3 \end{bmatrix}$$

And write an echelon matrix row equivalent to A.

Solution.

$$RankA = 1 + Rank = \begin{bmatrix} \begin{vmatrix} 1 & -2 \\ -1 & 0 \end{vmatrix} & \begin{vmatrix} 1 & 1 \\ -1 & 1 \end{vmatrix} & \begin{vmatrix} 1 & 0 \\ -1 & -2 \end{vmatrix} & \begin{vmatrix} 1 & 5 \\ -1 & 2 \end{vmatrix} \\ \begin{vmatrix} 1 & -2 \\ 1 & -6 \end{vmatrix} & \begin{vmatrix} 1 & 1 \\ 1 & 3 \end{vmatrix} & \begin{vmatrix} 1 & 0 \\ 1 & -2 \end{vmatrix} & \begin{vmatrix} 1 & 5 \\ 1 & 12 \end{vmatrix} \\ \begin{vmatrix} 1 & -2 \\ 2 & -3 \end{vmatrix} & \begin{vmatrix} 1 & 1 \\ 2 & 0 \end{vmatrix} & \begin{vmatrix} 1 & 0 \\ 2 & 2 \end{vmatrix} & \begin{vmatrix} 1 & 5 \\ 2 & 3 \end{vmatrix} \end{bmatrix}$$

$$= 1 + Rank \begin{bmatrix} -2 & 2 & -2 & 7 \\ -4 & 2 & -2 & 7 \\ 1 & -2 & 2 & -7 \end{bmatrix}$$

$$= 2 + Rank \begin{bmatrix} \begin{vmatrix} -2 & 2 \\ -4 & 2 \end{vmatrix} & \begin{vmatrix} -2 & -2 \\ -4 & 2 \end{vmatrix} & \begin{vmatrix} -2 & 7 \\ -4 & 7 \end{vmatrix} \\ \begin{vmatrix} -2 & 2 \\ 1 & -2 \end{vmatrix} & \begin{vmatrix} -2 & -2 \\ 1 & 2 \end{vmatrix} & \begin{vmatrix} -2 & 7 \\ 1 & -7 \end{vmatrix} \end{bmatrix}$$

$$2 + Rank \begin{bmatrix} 4 & -4 & 14 \\ 2 & -2 & 7 \end{bmatrix}$$

$$= 3 + Rank \begin{bmatrix} \begin{vmatrix} 4 & -4 \\ 2 & -2 \end{vmatrix} & \begin{vmatrix} 4 & 14 \\ 2 & 7 \end{vmatrix} \end{bmatrix}$$

$$= 3 + Rank \begin{bmatrix} 0 & 0 \end{bmatrix} = 3$$

An echelon matrix row equivalent to A is

$$\begin{bmatrix} 1 & -2 & 1 & 0 & 5 \\ 0 & -2 & 2 & -2 & 7 \\ 0 & 0 & 4 & -4 & 14 \\ 0 & 0 & 0 & 0 & 0 \end{bmatrix}$$

Example. Find the rank of the matrix

$$A = \begin{bmatrix} 2 & 1 & 3 & 5 \\ 15 & 8 & 1 & 12 \\ 11 & 5 & 8 & 6 \\ 12 & 8 & 7 & 10 \end{bmatrix}$$

Solution.

$$A = 1 + Rank \begin{bmatrix} 1 & -43 & -51 \\ -1 & -17 & -43 \\ 4 & -22 & -40 \end{bmatrix}$$

Rank

$$= 2 + Rank\begin{bmatrix} -60 & -94 \\ 150 & 164 \end{bmatrix} = 3 + Rank\begin{bmatrix} 4260 \end{bmatrix} = 4$$

An echelon matrix row equivalent to A is $\begin{bmatrix} 2 & 1 & 3 & 5 \\ 0 & 1 & -43 & -51 \\ 0 & 0 & -60 & -94 \\ 0 & 0 & 0 & 4260 \end{bmatrix}$

Proofs of Theorem

Proof of Corollary. If A nonsingular, then by $x = A^{-1} = A^{-1}0 = 0$ is the unique solution. Conversely, if the system $Ax = 0$ ha a unique solution $x = 0,$ then the columns a_j of A satisfy the equation $\sum_{j=1}^{n} x_i a_j = 0$.

Only when $x_j = 0, j = 1,2 ,\ldots, n i.e.,$ the columns of A are linearly independent and so A is nonsingular.

Definition (The solution space). The solution part is now obvious.

$x_1 - x_2 + 2x_3 + x_4 = 0$

$3x_1 + 2x_2 + x_4 = 0$

$4x_1 + x_2 + 2x_3 + 2x_4 = 0$

is given by

$$x_1 = -\frac{4}{5}x_3 - \frac{3}{5}x_4$$

$$x_2 = \frac{6}{5}x_3 + \frac{2}{5}x_4.$$

$x_3 = a\, and\;\; x_4 = b.$ Then

Let $x_3 = a \;\; and \;\; x_4 = b.Then$

$$\begin{bmatrix} x_1 \\ x_2 \\ x_3 \\ x_4 \end{bmatrix} = \begin{bmatrix} -\frac{4}{5}a-\frac{3}{5}b \\ \frac{6}{5}a+\frac{2}{5}b \\ a \\ b \end{bmatrix} = a\begin{bmatrix} -\frac{4}{5} \\ \frac{6}{5} \\ 1 \\ 0 \end{bmatrix} + b\begin{bmatrix} -\frac{3}{5} \\ \frac{2}{5} \\ 0 \\ 1 \end{bmatrix}$$

Thus every solution of the given system of equations is linear combinations of

$$\begin{bmatrix} -\frac{4}{5} \\ \frac{6}{5} \\ 1 \\ 0 \end{bmatrix} and \begin{bmatrix} -\frac{3}{5} \\ \frac{2}{5} \\ 0 \\ 1 \end{bmatrix}$$

It is easy to check that the set of all solutions forms a vector space, called the solution space of the given system

The vectors $\begin{bmatrix} -\frac{4}{5} \\ \frac{6}{5} \\ 1 \\ 0 \end{bmatrix} and \begin{bmatrix} -\frac{3}{5} \\ \frac{2}{5} \\ 0 \\ 1 \end{bmatrix}$ form a basis of this space.

Note that the two vectors are linearly independent.

Matrix of a Linear Transformation

Definition. Let V and W be finite dimensional vector space over the same field F with dim $V = n$ and dim $W = m$, Let $B = \{v_1, v_2, \ldots, v_n\}$ *and* $E = \{w_1, w_2, \ldots, w_m\}$ be any bases for V *and* W respectively. Any vector v in V can be expressed in a unique way as a linear combination of $v_1, v_2, \ldots, v_n$. Thus

$v = x_1v_1 + x_2v_2 + \ldots + x_nv_n, \quad x_i \in F, i = 1, 2, \ldots, n.$

We call $(x_1, x_2, \ldots, x_n)$ the coordinates vector of v relative to the basis B.

Let $T:V \to W$ be a linear transformation. The image T $(v_1), T(v_2), \dots, T(v_n)$ re elements of W and each can be expressed uniquely as a linear combination of the basis vectors $w_1, w_2, \dots, w_m$. Therefore,

$$T(v_1) = a_{11}w_1 + a_{21}w_2 + \dots + a_{m1}w_m$$

$$T(v_2) = a_{11}w_1 + a_{21}w_2 + \dots + a_{m1}w_m$$

$$\vdots \quad \vdots \quad \vdots \quad \dots \quad \vdots$$

$$T(v_j) = a_{1j}w_1 + a_{2j}w_2 + \dots + a_{mj}w_m$$

$$\vdots \quad \vdots \quad \vdots \quad \dots \quad \vdots$$

$$T(v_n) = a_{1n}w_1 + a_{2n}w_2 + \dots + a_{mn}w_m$$

Where $a_{ij} \in F$.

The $m \times n$ matrix whose jth column is the coordinate vector of $T(v_j)$ is called the matrix of T with respect to the basis $B\, and\, E$. Thus the matrix $A\, of\, T:V \to W$ relative to the basis $B\, and\, E\, is$

$$A = \begin{bmatrix} a_{11} & a_{12} \cdots & a_{ij} \cdots & a_{1n} \\ a_{21} & a_{22} \cdots & a_{2j} \cdots & a_{2n} \\ \vdots & \vdots \ \dots & \vdots & \vdots \\ a_{m1} & am_2 \cdots & a_{mj} \cdots & a_{mn} \end{bmatrix}$$

Conversely, If $A = [a_{ij}]\, is\, an\, m \times n$ matrix with entries $a_{ij} \in F, then\, A$ represent a linear transformation

$T: F^n \to F^m$

Defined by the equation

$T(x) = Ax$

$$x = \begin{bmatrix} x_1 \\ x_2 \\ \vdots \\ x_n \end{bmatrix}$$

Where is a column vector of F^n.

Example. Find the matrix of the linear transformation $T : R^3 \to R^4$ defined by

$T(x_1, x_2, x_3) = (x_1 + x_2, x_2 + x_3, x_1 - x_3, x_1)$

With respect to the standard bases for $R^3 \; and \; R^4$

Solution. Standard base B and E for $R^3 \; and \; R^4 \; are$

$B = \{(1,0,0),(0,1,0),(0,0,1)\} = \{e_1, e_2, e_3\}$

$E = \{(1,0,0,0,0),(0,1,0,0),(0,0,1,0),(0,0,0,1)\}$

$T(e_1) = (1,0,1,1)$

$= 1(1,0,0,0) + 0(0,1,0,0) + 1(0,0,1,0) + (0,0,0,1)$

$T(e_2) = (1,1,0,0)$

$= 1(1,0,0,0) + 1(0,1,0,0) + 0(0,0,1,0) + 0(0,0,0,1)$

$T(e_3) = (0,1,-1,0)$

$= 0(1,0,0,0,0) + 1(0,1,0,0) - 1(0,0,1,0) + 0(0,0,0,1).$

The matrix $A \, of \, T$ is

$$A = \begin{bmatrix} 1 & 1 & 0 \\ 0 & 1 & 1 \\ 1 & 0 & -1 \\ 1 & 0 & 0 \end{bmatrix}$$

Example. The matrix of a linear transformation $T : R^3 \to R^3$ is

$$A = \begin{bmatrix} 3 & 1 & 2 \\ 0 & 1 & 1 \\ -1 & 1 & 1 \end{bmatrix}$$

Determine T in terms of coordinates of a vector $x \in R^3$

Solution. Let $x = (x_1, x_2, x_3).T$ is given by

$$T(x) = Ax = \begin{bmatrix} 3 & 1 & 2 \\ 0 & 1 & 1 \\ -1 & 1 & 1 \end{bmatrix} \begin{bmatrix} x_1 \\ x_2 \\ x_3 \end{bmatrix} = \begin{bmatrix} 3x_1 + x_2 + 2x_3 \\ 0x_1 + x_2 + x_3 \\ -x_1 + x_2 + x_3 \end{bmatrix}$$

i.e.,
$$T \begin{bmatrix} x_1 \\ x_2 \\ x_3 \end{bmatrix} = \begin{bmatrix} 3x_1 + x_2 + 2x_3 \\ 0x_1 + x_2 + x_3 \\ -x_1 + x_2 + x_3 \end{bmatrix}$$

or $\quad T(x_1, x_2, x_3) = (3x_1 + x_2 + 2x_3, 0x_1 + x_2 + x_3, -x_1 + x_2 + x_3).$

SUMMARY AND DISCUSSION

Vector space from a very important topic in both pure and applied mathematics. Essentially it is a part of Linear Algebra In this chapter we discuss it and some of the concept replaced to it. Mainly we give a brief account of subspaces, their spanning sets, ; linear dependence and linear independence of subsets of a vector space, basic and dimension of a vector space, roe and column spaced, solution space of a linear system of equations, linear transformations and matrix of a linear transformation.

Definition. Let f be a field and V a nonempty set on whose elements an operation of addition is defined. Suppose that for every $a \in F$ and every $v \in V, av \, is \, an \, element \, of \, V$. Then V is called a **vector space** over F if the following axioms hold:

(i) $V \, is \, an \, abelian \, group \, under \, addition$

(ii) $a(v + w) = av + aw$

(iii) $(a + b)v = av + bv$

(iv) $a(bv)=(ab)v$

(v) $Iv=v,1$ being the multiplicative identity of F

$for\ all\ a,b\in F;v,w\in V.$

The expression $av\ ,a\in F,\ v\in V$ is called the scalar multiplicative of v by a. The elements of F are called scalars and the elements of V are called vectors. The vector space V over the frild F is sometimes written as $V(F).If\ the\ feild\ F,over\ which V\ is\ a$ vector space, is clearly understood form the context, then we simply write V instead of $V(F)$. A vector space V is called a real or complex vector space addition and field of real or complex numbers respectively.

Theorem. Let V be a vector space over a field F. then

(i) $a0=0,$ $for\ all\ a\in F$

(ii) $0v=0,$ $for\ all\ v\in V$

(iii) $(-a)v=a(-v)=-av,$ $for\ all\ a\in F,v\in V$

(iv) $if\ av=0\ then\ either\ a=0\ or\ v=0$ $for\ all\ a\in F,\ v\in V$

(v) $a(u-v)=au-av,$ $for\ all\ a\in F,u,v\in V.$

Subspaces

Definition. A nonempty subset W of a vector space V over a field F is said to be a subspace of V if W is itself a vector space over F under the same operations as defined in V.

Every vector space V has two subspaces, namely V Any other subspace of V is called a nontrivial subspace.

Theorem. Let V be a vector space over a field F and let W be a nonempty subset of V. Then W is a subspace of V if and only if.

$w_1,w_2\in W \quad \Rightarrow w_1+w_2\in W$

$w\in W,a\in F \quad \Rightarrow aw\in W$

In other words W is a subspace of V, if and only if, it is closed under the operations of addition and scalar multiplication as defined in V.

Corollary. A nonempty subset W of V is a subspace of V, if and only if, $w_1, w_2 \in W \; and \; a,b \in F$ imply $aw_1 + bw_2 \in W$.

Theorem. Let U and W be subspace of a vector space V over a field F. Then $U \cap W$ is also a subspace of V.

Theorem. The intersection of any number of subspace of a vector space V is a subspace proof V.

Definition. (Sum of two space). Let $U \; and \; W$ be two subspaces of a vector space V. Define

$$U + W = \{u + w : u \in U \; and \; w \in W\}$$

Clearly $U + W \subset V$.

We next show that $U + W$ is a subspace of V.

Theorem. if U, W are subspace of a vector space $V, \; then \; U + W$ is a subspace of $V \; contatining \; both \; U \; and \; W$. Further, $U + W$ is the smallest subspace containing both $U \; and \; W$.

Definition. $Let \, U, W$ be subspace of a vector space V. if $U \cap W = \{0\}$, then Thus a vector space V is called the direct sum of its subspace $U \; and \; W$ if

(i) $v = U + W$ and

(ii) $U \cap W = \{0 \cap\}$

Definition. Let V be a vector space over a field F and $v_1, v_2, \ldots v_m \in V$. Any vector in V of the form $a_1 v_1 + a_2 v_2 + \ldots + a_m v_m$,

where $a_i \in F \, (1 \le i \le m)$, is called a linear combination (over F) of $v_1, v_2, \ldots v_m$.

Definition. Let S be a nonempty subset of a vector space V over a field $F.$ The $\langle S\rangle$ is a subspace of V containing S and it is the smallest subspace of V containing $S.$

If W is any other subspace V containig $S,$ then it contain all vectors of the from $\sum_{i=1}^{k} a_i v_i,$ where $a_i \in F, v_i \in V \; and \; K$ is natural numbers. Hence $\langle S\rangle W.$ Thus $\langle S\rangle$ is the smallest subspace of V containing $S.$ $\langle S\rangle$ is said to be spanned (or generated) by $S \, and \, S$ is called the spanning set for $\langle S\rangle.$

Corollary. $if \; S, T$ are subsets of V then
$S \subset T \quad implies \langle S\rangle \subset \langle T\rangle.$

Definition. A vector space V is said to be finite dimensional if there is a finite subset $S \, in V$ such that $V = \langle S\rangle.$

A vector space which is not finite dimensional is called infinite dimensional. Thus $P(x)$ is an infinite dimensional vector space.

Definition. $Let V$ be a vector spave over a field $F.$ The vector $v_1, v_2 \ldots, v_m m \in V$ are said to be linearly dependent over F if
$av_1 + a_2 v_2 + \ldots + a_m v_m = 0 \, implies \; each \, a_1 = 0, i-1, 2, \, , m.$

In this case we say that $\{v_1, v_2, \ldots, v_m\}$ is a linearly independent set.

An infinite set S of vectors is linearly independent if each of its finite subsets is linearly independent. Otherwise it is linearly dependent.

The empty set $\varnothing$ is defined to be linearly independent.

Theorem. (i) A Set $S = \{v_1, v_2, \ldots, v_n\} of \, n$ vectors $(n \geq 2)$ in a vector space V is linearly dependent if and only if at least one of the vectors in S is a linear combination of the remaining vectors of the set

(ii) A set $S = \{v_1, v_2, \ldots, v_n\}$ of vector $n \geq 2,$ in a vector space V is linearly dependent if and only if one of the vectors in S is a linear combination of the vectors preceding it.

Definition. A set of linearly independent vectors spanning a vector space V is called a basis for V. Unless $V = \{0\},$ the basis vectors are nonzero, since we can ignore any zero vector from a spanning set.

Any finite dimensional vector space contain basis.

The dimension of a vector space V over a field $F,$ written dim $V,$ is the number of elements in a basis for V.

All basis of a finite dimensional vector space contain the same number of elements.

Let V be a space such that dim $V = n < \infty, A$ set of vectors $\{v_1, v_2, \ldots, v_n\} \subset V$ is a basis for V if and only if each vectors in V is uniquely expressible as a linear combination of $v_1, v_2, \ldots, v_n$.

Let $S = \{v_1, v_2, \ldots, v_n\}$ be a basis for $n = dimensional$ vector space V over a field $F.$ Then every set with more than n vectors is linearly dependent.

Let $v_1, v_2, \ldots, v_n$ be linearly independent in a vector space V over a field $F.$ if v is any nonzero vector in $V,$ then the set $\{v_1, v_2, \ldots, v_n, \text{v}\}$ is a linearly independent if and only if v is not in the linear span $\langle v_1, v_2, \ldots, v_n \rangle$

.

Let V be an n dimensional vector spacr over a field $F.$ Then every set $S = \{v_1, v_2, \ldots, v_n\}$ *of* n linearly independent vectors in V is a basis for $V.$

Any linearly Independent set f vectors in a finite dimensional vector space V can be extended to a basis for $V.$

If W is a subspace of a finite dimension vectors space $V,$ then dim $W \leq \dim V$ *Moreover,* if dim $W = \dim V, then\, W = V.$

A vector space V is the direct sum of its subspace $U\, and\, W$ if and only if each $v \in V$ can be written uniquely as $(i)\ v = u + w,\ \ u \in U,\ w \in W.$

Theorem. If $U\, and\, W$ are finite dimensional subspace of a vector space V over a field $F,$ then

(i) $\dim(U+W) = deim U + \dim W - \dim(U \cap W).$

(ii) if $U \cap W = \{0\}\ then V = U \oplus W$ and $\dim V = \dim U + \dim W.$

Definition. Let $A = \left[a_{ij}\right]$ be an $m\,xn$ matrix over $R, m \leq n.$ As defined in row of A

$$P_1 = \left[a_{11}\, a_{12} \ldots a_{1n}\right],\ p_2 = \left[a_{21}\, a_{22} \ldots a_{2n}\right], \ldots,$$

$$P_m = \left[am_1\ \ am_2 \ldots a_{mn}\right]$$

are called row vectors of A. The subspace $\boldsymbol{R}^n$ spanned by $\{p_1, p_2, \ldots, p_m\}$ is called the row space of the matrix $A.$

Similarly , the subspace spanned by the column vectors

$$\begin{bmatrix} a_{11} \\ a_{21} \\ \vdots \\ a_{m1} \end{bmatrix}, \begin{bmatrix} a_{12} \\ a_{22} \\ \vdots \\ a_{mn} \end{bmatrix}, \ldots, \begin{bmatrix} a_{1n} \\ a_{2n} \\ \vdots \\ a_{mn} \end{bmatrix}$$

Of A is called the column space of $A.$ The dimension of the row space is called the row rank of the matrix A. Thus the row rank is the maximum number of linearly independent row vectors of A, similarly ,the column rank of A is the maximum number of linearly independent column vector of A.

Theorem. Row equivalent matrices have the same row space.

Row equivalent matrices have the same row rank.

If P is nonsingular, then the matrices A and PA have the same row rank.

The row rank of a nonsingular matrix A of order $n\,x\,n\,is\,n.$

The rank of a matrix A was found by reducing it to on echelon form. To reduce A to an echelon from many operations are needed and in the process fractions creep in which make the computations awkward and cumbersome. The Alternative Method is very elegant involving no fractions and it also yields an echelon matrix that is row equivalent to A and whose nonzero rows constitute a basis for the row space of A.

CHAPTER 5

APPLICATIONS OF REAL VECTOR SPACES

INTRODUCTION

In this section we will discuss another factorization of a matrix called the QR- Factorization of A. This type of factorization is widely used in computer codes to find the eigenvalue of a matrix, to solve linear systems, and to find least squares approximations.

Theorem. If A is an $m \times n$ matrix with linearly independent columns, then A can be factored as $A = QR,$ where Q is an $m \times n$ matrix whose columns form an othonormal basis for the column space of $A and R is an n \times n$ nonsingular upper triangular matrix.

Proof. Let $u_1, u_2, \ldots, u_n$ denote the linearly independent columns of $A,$ which form a basis for the column space of $A.$ By using the Gram-Schmidt process, we can obtain an orthonormal basis $w_1, w_2, \ldots, w_n$ for the column space of $A.$ Recall how this orthonormal basis was obtained. We first constructed an orthonormal basis $v_1, v_2, \ldots, v_n$ as follows:

$$v_i = u_i - \frac{u_i . v_1}{v_1 . v_1} v_1 - \frac{u_i . v_2}{v_2 . v_2} v_2 - \ldots - \frac{u_i . v_{i-1}}{v_{i-1} . \ v_{i-1}} v_{i-1} \qquad (1)$$

Finally, $w_i = \dfrac{1}{v_i} v_i \ for\, i = 1, 2, 3, \ldots, n.$ Now each of the vectors u_i can be written as a linear combination of the $w-vectors:$

$$u_1 = r_{11}w_1 + r_{21}w_2 + \ldots + r_{n1}w_n$$

$$u_2 = r_{12}w_1 + r_{22}w_2 + \ldots + r_{n2}w_n \qquad (2)$$

$$u_n = r_{1n}w_1 + r_{2n}w_2 + \ldots + r_{nn}w_n$$

Since we know that $r_{ji} = u_i.w_j$

Moreover, from equation (1), we see that u_i lies in $span\{v_1, v_2, \ldots v_i\} = span\{w_1, w_2, \ldots, w_i\}$.

Since w_j is orthogonal to $span\{w_1, w_2, \ldots, w_i\}\ for\ j > i,$ it is orthogonal to u_i. Hence $r_{ji} = 0\ for\ j > i.$ Let Q be the matrix whose columns are $w_1, w_2, \ldots, w_j$. Let

$$r_j = \begin{bmatrix} r_1 j \\ r_{2j} \\ \vdots \\ r_{nj} \end{bmatrix}$$

Then the equations in (2) can be written in matrix form as $A = [u_1\ \ u_2\ \ldots\ u_n] = [Qr_1\ \ Qr_2\ \ \ldots\ \ Qr_n] = QR.$

Where R is the matrix whose columns are $r_1, r_2, \ldots r_n$. Thus

$$R = \begin{bmatrix} & r_{11} & r_{12} & \cdots & & r_{1n} \\ & 0 & & r_{22} & \cdots & r_{2n} \\ 0 & & 0 & & \cdots & \\ & \vdots & \vdots & \vdots & \vdots & \\ 0 & & 0 & & \cdots & r_{nn} \end{bmatrix}$$

We now show that R is nonsingular. Let x be a solution to the linear system $Rx = 0$. Multiplying this equation by Q on the left. We have $Q(Rx) = (QR)x = Ax = Q0 = 0.$

As we know that the homogeneous system $Ax = 0$ can be written as $x_1u_1 + x_2u_2 + \ldots + x_nu_n = 0$.

Where $x_1, x_2, \ldots, x_n$ are the components of the vector x. Since the columns of A are linearly independent,

$x_1 = x_2 = \ldots = x_n = 0$

So x must be the zero vector. Then R is nonsingular.

Example. Find the QR- Factorization of

$$A = \begin{bmatrix} 1 & -1 & -1 \\ 1 & 0 & 0 \\ 1 & -1 & 0 \\ 0 & 1 & -1 \end{bmatrix}$$

Solution. Letting the column of A be denoted by u_1, u_2 *and* u_3. We let W be the subspace of R^4 with u_1, u_2, u_3 as a basis. Applying the Gram – Schmidt process to this basis, starting with u_1, we find the following orthonormal basis for the column space of A:

$$w_1 = \begin{bmatrix} \frac{1}{\sqrt{3}} \\ \frac{1}{\sqrt{3}} \\ \frac{1}{\sqrt{3}} \\ 0 \end{bmatrix} \quad w_2 = \begin{bmatrix} -\frac{1}{\sqrt{15}} \\ \frac{2}{\sqrt{15}} \\ -\frac{1}{\sqrt{15}} \\ \frac{3}{\sqrt{15}} \end{bmatrix} \quad w_3 = \begin{bmatrix} -\frac{4}{\sqrt{35}} \\ \frac{3}{\sqrt{35}} \\ \frac{1}{\sqrt{35}} \\ -\frac{3}{\sqrt{35}} \end{bmatrix}$$

Then

$$Q=\begin{bmatrix} \frac{1}{\sqrt{3}} & -\frac{1}{\sqrt{15}} & -\frac{4}{\sqrt{35}} \\ \frac{1}{\sqrt{3}} & \frac{2}{\sqrt{15}} & \frac{3}{\sqrt{35}} \\ \frac{1}{\sqrt{3}} & -\frac{1}{\sqrt{15}} & \frac{1}{\sqrt{35}} \\ 0 & \frac{3}{\sqrt{15}} & -\frac{3}{\sqrt{35}} \end{bmatrix} \approx \begin{bmatrix} 0.5774 & -0.2582 & -0.6761 \\ 0.5774 & 0.5164 & 0.5071 \\ 0.5774 & -0.2582 & 0.1690 \\ 0 & 0.7746 & -0.5071 \end{bmatrix}$$

And $R=\begin{bmatrix} r_{11} & r_{12} & r_{13} \\ 0 & r_{22} & r_{23} \\ 0 & 0 & r_{33} \end{bmatrix}$.

Where $r_{ji}=u_i.w_j$. Thus

$$R=\begin{bmatrix} \frac{3}{\sqrt{3}} & -\frac{2}{\sqrt{3}} & -\frac{1}{\sqrt{3}} \\ 0 & \frac{5}{\sqrt{15}} & -\frac{2}{\sqrt{15}} \\ 0 & 0 & \frac{7}{\sqrt{35}} \end{bmatrix} \approx \begin{bmatrix} 1.7321 & -1.1547 & -0.5774 \\ 0 & 1.2910 & -0.5164 \\ 0 & 0 & 1.1832 \end{bmatrix}$$

Now we can verify $A=QR.$

Least Squares

We recall that an $m \times n$ linear system $Ax=b$ is inconsistent if it has no solution. We also know that $Ax=b$ is consistent if and only if b belongs to the column space of $A.$ Equivalently, $Ax=b$ is inconsistent if and only if b is not in the column space of $A.$ Inconsistent systems do indeed arise in many situations and we must determine how to deal with them. Our approach is to change the problem so that we do not require that the matrix equation $Ax = b$ be satisfied. Instead, we seek a vector $\hat{x}\, in\, R^n$ such that $A\hat{x}$ is as close to b as possible. If W is the column space of $A,$ then from theorem, it follows that the vector in W that is closed to b is $proj_w b.$ That is $b - w,$ for w in W, is minimized when $w = proj_w b.$ Thus, if we find $\hat{x}$ such that $A\hat{x} = proj_w b,$ then we assured that $b - A\hat{x}$ will be as small as possible. A we already know from previous theorem $b - proj_w b = b - A\hat{x}$ is orthogonal to each column of A. In terms of a matrix equation we have

$$A^T\left(A\hat{x} - b\right) = 0$$

or, equivalently, $A^T A\hat{x} = A^T b.$

Thus $\hat{x}$ is a solution to $A^T Ax = A^T b.$ (1)

Any solution to (1) is called a least squarcs solution to the linear system $Ax = b.$ Equation (1) is called the normal system of equations associated with $Ax = b,$ or just the normal system. Observe that if A is nonsingular, a least squares solution to $Ax = b$ is just the usual solution $x = A^{-1}b.$

To compute a least squares solution $\hat{x}$ to the linear system $Ax = b,$ we can proceed as follows. Let $\{w_1, w_2, \ldots, w_m\}$ be an orthonormal basis for the column space $W\, of\, A.$ Then equation (1) yields

$$proj_w b = \left(b.w_1\right)w_1 + \left(b.w_2\right)w_2 + \ldots + \left(b.w_m\right)w_m.$$

Recall that to find an orthonormal basis for W we first find a basis for W by transforming A^T to reduce row echelon form and then taking the transposes of the nonzero rows as a basis for $W.$ Next apply the Gram- Schmidt process to this basis to find an orthonormal basis for $W.$ The procedure just outlined is theoretically valid when we assume that exact arithmetic is used. However, even small numerical errors, due to, say round off, may adversely affect the results. Thus more sophisticated algorithms are required for numerical applications.

Theorem. If $A \, is \, an \, m \times n$ matrix with $rank\ A = n,$ then $A^T A$ is nonsingular and the linear system $Ax = b$ has a unique least squares solution given by $\hat{x} = \left(A^T A \right)^{-1} A^T b.$ That is, the normal system of equations has a unique solution.

Proof. If A has rank $n,$ then the columns of A are linearly independent. The matrix $A^T A$ is a nonsingular provided the linear system $A^T Ax = 0$ has only the zero solution. Multiplying both sides of $A^T Ax = 0$ by x^T gives

$$0 = x^T A^T Ax = \left(Ax \right)^T Ax = \left(Ax \right)^T \left(Ax \right) = \left(Ax \right).\left(Ax \right).$$

It then follows from theorem that $Ax = 0.$ But this implies that we have a linear combination of the linearly independent columns of A that is zero, hence $x = 0.$ Thus $A^T A$ is nonsingular and equation (1) has the unique solution $\hat{x} = \left(A^T A \right)^{-1} A^T b.$

Example. Determine a least squares solution to $Ax = b,$ where

$$A = \begin{bmatrix} & & & 3 \\ & & & 2 \\ 1 & 2 & -1 & 1 \\ 2 & 1 & 1 & 0 \\ -2 & 3 & 4 & 3 \\ 4 & 2 & 1 & 0 \\ 0 & 2 & 1 & \\ 1 & -1 & 2 & \end{bmatrix} \qquad b = \begin{bmatrix} \\ \\ 1 \\ 5 \\ -2 \\ 1 \\ 3 \\ 5 \end{bmatrix}.$$

Solution. Using row reduction, we can show that $rank\ A = 4.$ We now form the normal system $A^T Ax = A^T b,$

$$\begin{bmatrix} 26 & 5 & -1 & 5 \\ 5 & 23 & 13 & 17 \\ -1 & 13 & 24 & 6 \\ 5 & 17 & 6 & 23 \end{bmatrix} x = \begin{bmatrix} 24 \\ 4 \\ 10 \\ 20 \end{bmatrix}$$

By theorem, the normal system has a unique solution. Applying Gauss- Jordan reduction, we have the unique least squares solution.

$$\hat{x} \approx \begin{bmatrix} 0.9990 \\ -2.0643 \\ 1.1039 \\ 1.8902 \end{bmatrix}$$

If W is the column space of $A,$ then

$$proj_w b = A\hat{x} \approx \begin{bmatrix} 1.4371 \\ 4.8181 \\ -1.8852 \\ 0.9713 \\ 2.6459 \\ 5.2712 \end{bmatrix}.$$

which is the vector in W such that $b - w$, w in W, is minimized. That is, $\min_{w \, in \, W} b - w = b - A\hat{x}$.

When A is an $m \times n$ matrix whose rank is n, it is computationally more efficient to solve equation (1) by Gauss- Jordan reduction than to determine $\left(A^T A\right)^{-1}$ and then form the product $\left(A^T A\right)^{-1} A^T b.$ An even better approach is to use the QR- factorization of A.

Least Square using QR – Factorization

Let $A = QR$ be a QR – Factorization of A. Substituting this expression for A into equation (1), we obtain

$$(QR)^T (QR)x = (QR)^T b$$

Or $R^T \left(Q^T Q\right) Rx = R^T Q^T b.$

Since the columns of Q form an orthonormal set, we have $Q^T Q = I_m$, so

$$R^T Rx = R^T Q^T b.$$

Since R^t is a nonsingular matrix, we obtain

$$Rx = Q^T b.$$

Using the fact that R is upper triangular, we easily solve this linear system by back substitution to obtain $\hat{x}$.

Example. Solve above example using the QR- Factorization of A.

Solution. We use the Gram-Schmidt process, carrying out all computations in any software like MATLAB. We find that Q is given by

$$Q = \begin{bmatrix} & & & 0.3409 \\ & & & 0.4244 \\ -0.1961 & -0.3851 & 0.5099 & -0.2177 \\ -0.3922 & -0.1311 & -0.1768 & -0.5076 \\ 0.3922 & -0.7210 & -0.4733 & 0.4839 \\ -0.7845 & -0.2622 & -0.1041 & 0.4055 \\ 0 & -0.4260 & 0.0492 & \\ -0.1961 & 0.2540 & -0.6867 & \end{bmatrix}$$

And R is given by

$$R = \begin{bmatrix} -5.0990 & -0.9806 & 0.1961 & -0.9806 \\ 0 & -4.6945 & -2.8102 & -3.4164 \\ 0 & 0 & -4.0081 & 0.8504 \\ 0 & 0 & 0 & 3.1054 \end{bmatrix}$$

$$Q^T b = \begin{bmatrix} 4.7068 \\ -0.1311 \\ 2.8172 \\ 5.8699 \end{bmatrix}$$

Then

Finally, solving $Rx = Q^T b$

$$x = \begin{bmatrix} 0.9990 \\ -2.0643 \\ 1.1039 \\ 1.8902 \end{bmatrix}.$$

We find

Which agrees with $\hat{x}$ as obtained in earlier example.

Example. In the manufacture of product A, the amount of the compound beta present in the product is controlled by the amount of the ingredient alpha used in the process. In manufacturing a gallon of A, the amount of alpha used and the amount of beta present are recorded. The following data were obtained:

Alpha Used	3	4	5	6	7	8	9	10	11	12
Beta Present	4.5	5.5	5.7	6.6	7.0	7.7	8.5	8.7	9.5	9.7

Suppose it is assumed that the relationship between the amount of alpha used and the amount of beta present is given by a linear equation so that the graph is a straight line. It would thus not be reasonable to connect the plotted points by drawing a curve that passes through every point. Moreover, the data are of a probabilistic nature. That is, they are not deterministic in the sense that if we repeated the experiment, we would expect slightly different values of beta for the same values of alpha. Since all measurements are subject to experimental error. Thus the plotted points do not lie exactly on a straight line. We shall now use the method of least squares to obtain the straight line that best fits the given data. This line is called the least square line.

Suppose that we are given n points $(x_1, y_1),(x_2, y_2),\ldots,(x_n, y_n)$ where at least two of the x_i are distinct. We are interested in finding the least squares line

$$y = b_1 x + b_0 \tag{2}$$

That best fits the data. If the points $(x_1, y_1),(x_2, y_2),\ldots,(x_n, y_n)$ were exactly on the least square line, we would have

$$y_i = b_1 x_i + b_0 \tag{3}$$

Since some of these points may not lie exactly on the line, we have

$$y_i = b_1 x_i + b_0 + d_i, \qquad i = 1, 2, \ldots, n,. \tag{4}$$

Where d_i is the vertical deviation from the point (x_i, y_i) to the desired least squares line. The quantity d_i can be positive , negative or zero. If we let

$$b = \begin{bmatrix} y_1 \\ y_2 \\ \vdots \\ y_n \end{bmatrix}, A = \begin{bmatrix} x_1 & 1 \\ x_2 & 1 \\ \vdots & \vdots \\ x_n & 1 \end{bmatrix}, x = \begin{bmatrix} b_1 \\ b_0 \end{bmatrix}$$

Then we may write the n equations in (4) as a single matrix equation

$$b = Ax + d$$

Since the linear system $Ax = b$ is usually inconsistent. We find a least square solution $\hat{x}$ to $Ax = b$. Recall that at least two of the x_i are distinct. So $rank\,A = 2$.

Then by theorem it implies that $Ax = b$ has a unique least squares solution given by

$\hat{x} = \left(A^T A\right)^{-1} A^T b.$ We are then assured that $A\hat{x}$ will be as close as possible.

For each $x_i = 1, 2, \ldots, n.$ Let $\widehat{y_i} = b_1 x_1 + b.$ Then $\left(x_i, \widehat{y_i}\right)$ are the points on the least square line. Using formula from previous sections it follows that d is minimized. Equivalently,

$d^2 = d_1^2 + d_2^2 + \ldots + d_n^2$ will be minimized.

Procedure for finding the least square line $y = b_1 x + b_0$ for the data $(x_1, y_1), (x_2, y_2), \ldots, (x_n, y_n),$ where at least two of the x_i are different, is as follows.

Step I. Let

$$b = \begin{bmatrix} y_1 \\ y_2 \\ \vdots \\ y_n \end{bmatrix}, \quad A = \begin{bmatrix} x_1 & 1 \\ x_2 & 1 \\ \vdots & \vdots \\ x_n & 1 \end{bmatrix}, \quad x = \begin{bmatrix} b_1 \\ b_0 \end{bmatrix}$$

Step II. Solve the normal system

$A^T A x = A^T b$

For x by Gauss- Jordan reduction.

Example

i) Find an equation of the least squares lines for the data of previous example.

ii) Use the equation obtained in part (i) to predict the number of ounces of beta present in a gallon of product A if 30 ounces of alpha are used per gallon.

Solution

We have

$$b=\begin{bmatrix}4.5\\5.5\\5.7\\6.6\\7.0\\7.7\\8.5\\8.7\\9.5\\9.7\end{bmatrix}, A=\begin{bmatrix}3&1\\4&1\\5&1\\6&1\\7&1\\8&1\\9&1\\10&1\\11&1\\12&1\end{bmatrix}, x=\begin{bmatrix}b_1\\b_0\end{bmatrix}.$$

Then $A^T A=\begin{bmatrix}645&75\\75&10\end{bmatrix} and\ A^T b=\begin{bmatrix}598.6\\73.4\end{bmatrix}.$

Solving the normal system $A^T Ax = A^T b$ by Gauss- Jordan reduction we obtain

$$x=\begin{bmatrix}b_1\\b_0\end{bmatrix}=\begin{bmatrix}0.583\\2.967\end{bmatrix}$$

Then $b_1=0.583\ \ and\ b_0=2.967.$

Hence, an equation for the least square line

$y=0.583x+2.967$ (5)

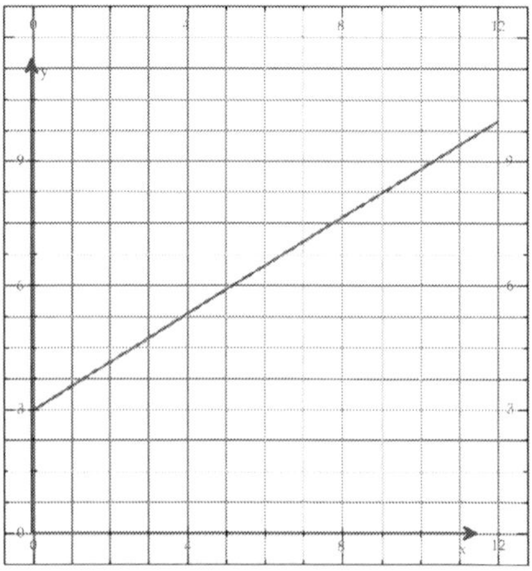

Where y is the amount of beta present and x is the amount of alpha used.

ii) If $x = 30,$ then substituting in (5), we obtain

$y = 20.457.$

Thus there would be 20.457 ounces of beta present in a gallon of A.

Example. The following data show atmospheric pollutants y_i at half-hour intervals t_i.

t_i	1	1.5	2	2.5	3	3.5	4	4.5	5
y_i	−0.15	0.24	0.68	1.04	1.21	1.15	0.86	0.41	−0.08

A plot of these data points as shown in figure, suggest that a quadratic polynomial

$y = a_2t^2 + a_1t + a_0$ may produce a good model for these data.

Now we, have

$$A = \begin{bmatrix} 1 & 1 & 1 \\ 2.25 & 1.5 & 1 \\ 4 & 2 & 1 \\ 6.25 & 2.5 & 1 \\ 9 & 3 & 1 \\ 12.25 & 3.5 & 1 \\ 16 & 4 & 1 \\ 20.25 & 4.5 & 1 \\ 25 & 5 & 1 \end{bmatrix}, \quad x = \begin{bmatrix} a_2 \\ a_1 \\ a_0 \end{bmatrix}, b = \begin{bmatrix} -0.15 \\ 0.24 \\ 0.68 \\ 1.04 \\ 1.21 \\ 1.15 \\ 0.86 \\ 0.41 \\ -0.08 \end{bmatrix}$$

The rank of A is 3, and the normal system is

$$\begin{bmatrix} 1583.25 & 378 & 96 \\ 378 & 96 & 27 \\ 96 & 27 & 9 \end{bmatrix} \begin{bmatrix} a_2 \\ a_1 \\ a_0 \end{bmatrix} = \begin{bmatrix} 54.65 \\ 16.71 \\ 5.36 \end{bmatrix}$$

Applying Gauss- Jordan reduction we obtain

$$\hat{x} \approx \begin{bmatrix} -0.3274 \\ 2.\ 0067 \\ -1.9317 \end{bmatrix}$$

So we obtain the quadratic polynomial model

$y = -0.3247t^2 + 2.0067t - 1.9317.$

Figure shows that the data set indicated with + and the graph of y. We see that y is close to each data point but is not required to go through the data.

SUMMARY AND DISCUSSION

In this chapter we discussed the application of real vector space in different fields with some specific problems. Using different methods specially QR algorithm, Method of Least square and extended form of QR methods.

Theorem. If A is an matrix with linearly independent columns, then A can be factored as $A = QR,$ where Q, is an $m \times n$ matrix whose columns form an orthonormal basis for the column space of $A \, and \, R \, is \, an \, n \times n$ nonsingular upper triangular matrix.

CHAPTER 6

EIGEN VALUES, EIGEN VECTORS, AND DIAGONALIZATION

INTRODUCTION

In this chapter we will discuss about Eigen Values, Eigen Vectors, and Diagonalization. For this chapter every matrix considered is a square matrix. Let A be an $n \times n$ matrix. Then, as we have seen in previous chapter, the function $L: R^n \to R^n$ defined by $L(x) = Ax, for\, x\, in\, R^n$, is a linear transformation. A question of considerable importance in a great many applied problems is the determination of vector x, if there are any, such that $x\, and\, Ax$ are parallel. Such questions arise in all applications involving vibrations; chemical engineering, biology, differential equations and so on. In this section we shall formulate this problem precisely; we also define some pertinent terminology. In the next section we solve this problem for symmetric matrices and briefly discuss the situation in the general case.

Definition. Let $A\, be\, an\, n \times n$ matrix. The real number λ is called an eigenvalue of A if there exists a nonzero vector x in R^n such that

$$Ax = \lambda x. \qquad (1)$$

Every non zero vector x satisfying (1) is called an eigenvector of A associated with the eigenvalue λ. We might mention that the word "eigenvalue" is a hybrid one ("eigen" in German means "proper"). Eigen values are also called proper values, characteristic values, and latent values; and eigen vectors are also called proper vectors, and so on, accordingly.

Note that $x = 0$ always satisfies (1), but 0 is not an eigen vector, since we insist that an eigen vector be a nonzero vector

In some applications one encounters matrices with complex entries and vector spaces with scalars that are complex numbers. In such a setting the preceding definition of eigenvalue is modified so that an eigen value can be real or a complex number (in some special cases otherwise, we require that an eigen value be a real number).

Example. If A is the identity matrix I_n, then the only eigenvalue is $\lambda = 1$; every nonzero vector in R^n is an eigenvector of A associated with the eigen value $\lambda = 1$;

$I_n x = 1x.$

Example. Let $A = \begin{bmatrix} 0 & \frac{1}{2} \\ \frac{1}{2} & 0 \end{bmatrix}.$

Then $A\begin{bmatrix} 1 \\ 1 \end{bmatrix} = \begin{bmatrix} 0 & \frac{1}{2} \\ \frac{1}{2} & 0 \end{bmatrix}\begin{bmatrix} 1 \\ 1 \end{bmatrix} = \begin{bmatrix} \frac{1}{2} \\ \frac{1}{2} \end{bmatrix} = \frac{1}{2}\begin{bmatrix} 1 \\ 1 \end{bmatrix}.$

So that $x_1 = \begin{bmatrix} 1 \\ 1 \end{bmatrix}$ is an eigenvector of A associated with the eigenvalue $\lambda_1 = \frac{1}{2}.$ Also,

$$A\begin{bmatrix} 1 \\ -1 \end{bmatrix} = \begin{bmatrix} 0 & \frac{1}{2} \\ \frac{1}{2} & 0 \end{bmatrix}\begin{bmatrix} 1 \\ -1 \end{bmatrix} = \begin{bmatrix} -\frac{1}{2} \\ \frac{1}{2} \end{bmatrix} = -\frac{1}{2}\begin{bmatrix} 1 \\ -1 \end{bmatrix}.$$

So that $x_2 = \begin{bmatrix} 1 \\ -1 \end{bmatrix}$ is an eigenvector of A associated with the eigenvalue $\lambda_2 = -\frac{1}{2}.$

An eigenvalue λ of A can have associated with it many different eigen vectors. In fact, if x is an eigenvector of A associated with $\lambda\left(i.e., Ax = \lambda x\right)$ and r is any nonzero real number, then

$A(rx)=r(Ax)=r(\lambda x)=\lambda(rx).$

Thus rx is also an eigenvector of A associated with λ.

Example. Let $A=\begin{bmatrix}0 & 0\\0 & 1\end{bmatrix}$ then

$$A\begin{bmatrix}1\\0\end{bmatrix}=\begin{bmatrix}0 & 0\\0 & 1\end{bmatrix}\begin{bmatrix}1\\0\end{bmatrix}=\begin{bmatrix}0\\0\end{bmatrix}=0\begin{bmatrix}1\\0\end{bmatrix}.$$

So that $x_1=\begin{bmatrix}1\\0\end{bmatrix}$ is an eigenvector of A associated with the eigenvalue $\lambda_2=1$.

Above example points out the fact that although the zero vector, by definition, cannot be an eigenvector, the number zero can be an eigen value.

Computing Eigen values and Eigen vectors

Thus far we have found the eigen values and associated eigen vectors of a given matrix by inspection, geometric arguments, or very simple algebraic approaches. In the following example, we compute the eigen values and associated eigen vectors of a matrix by a somewhat more systematic method.

Example. Let $A=\begin{bmatrix}1 & 1\\-2 & 4\end{bmatrix}$

We wish to find the eigen values of A and their associated eigenvectors. Thus we wish to find all real numbers λ and all nonzero vectors

$$x=\begin{bmatrix}x_1\\x_2\end{bmatrix}$$

Satisfying (1), that is,

$$\begin{bmatrix}1 & 1\\-2 & 4\end{bmatrix}\begin{bmatrix}x_1\\x_2\end{bmatrix}=\lambda\begin{bmatrix}x_1\\x_2\end{bmatrix} \qquad (2)$$

Equation (2) becomes

$x_1+x_2=\lambda x_1$

$$-2x_1 + 4x_2 = \lambda x_2,$$

Or

$$(\lambda - 1)x_1 - x_2 = 0$$

$$2x_1 + (\lambda - 4)x_2 = 0 \qquad (3)$$

Equation (3) is a homogeneous system of two equations in two unknowns. From Corollary, it follows that the homogeneous system in (3) has a nontrivial solution if and only if the determinant of its coefficient matrix is zero; that is, if and only if

$$\begin{vmatrix} \lambda - 1 & -1 \\ 2 & \lambda - 4 \end{vmatrix} = 0.$$

This means that

$(\lambda - 1)(\lambda - 4) + 2 = 0,$ or

$$\lambda^2 - 5\lambda + 6 = 0$$

$$(\lambda - 3)(\lambda - 2) = 0$$

Hence $\lambda_1 = 2\, and\, \lambda_2 = 3$ are the eigen values of A. To find all eigen vectors of A associated with $\lambda_1 = 2,$ we form the linear system $Ax = 2x.$

or $\begin{bmatrix} 1 & 1 \\ -2 & 4 \end{bmatrix}\begin{bmatrix} x_1 \\ x_2 \end{bmatrix} = 2\begin{bmatrix} x_1 \\ x_2 \end{bmatrix}.$

This gives

$$x_1 + x_2 = 2x_1$$

$$-2x_1 + 4x_2 = 2x_2$$

or $(2 - 1)x_1 - x_2 = 0$

$$2x_1 + (2 - 4)x_2 = 0$$

or $x_1 - x_2 = 0$

$$2x_1 - 2x_2 = 0$$

Note that we could have obtained this last homogeneous system by merely substituting $\lambda = 2$ in (3). All solutions to this last system are given by

$$x_1 = x_2$$

$$x_2 = any\, real\, number\, r.$$

Hence all eigenvectors associated with the eigenvalue $\lambda_1 = 2$ are given by $\begin{bmatrix} r \\ r \end{bmatrix}$, r any nonzero real number. In particular,

$x_1 = \begin{bmatrix} 1 \\ 1 \end{bmatrix}$ is an eigenvector associated with $\lambda_1 = 2.$ Similarly, for $\lambda_2 = 3$ we obtain from (3),

$$(3-1)x_1 - x_2 = 0$$

$$2x_1 + (3 \quad 4)x_2 - 0$$

Or $2x_1 - x_2 = 0$

$$2x_1 - x_2 = 0.$$

All solutions to this last homogeneous system are given by

$$x_1 = \frac{1}{2}x_2$$

$$x_2 = any\, real\, number\, r.$$

Hence all eigen vectors associated with the eigen value $\lambda_2 = 3$ are given by $\begin{bmatrix} \frac{1}{2}r \\ r \end{bmatrix}, r$ any nonzero real number. In particular,

$x_2 = \begin{bmatrix} 1 \\ 2 \end{bmatrix}$ is an eigen vector associated with the eigen value $\lambda_2 = 3.$

In above examples, we found eigen values and eigen vectors by inspection, whereas in last example, we proceed in a more systematic fashion. We use the procedure as our standard method.

Definition. Let $A=\left[a_{ij}\right]$ be an $n \times n$ matrix. The determinant

$$f(\lambda)=\det(\lambda I_n - A)=\begin{vmatrix} \lambda - a_{11} & -a_{12} & \cdots & -a_{1n} \\ -a_{21} & \lambda - a_{22} & \cdots & -a_{2n} \\ \vdots & \vdots & \vdots & \vdots \\ -a_{n1} & -a_{n2} & \cdots & \lambda - a_{nn} \end{vmatrix} \qquad (4)$$

is called the characteristic polynomial of A. The equation $f(\lambda)=\det(\lambda I_n - A)=0$ is called the characteristic equation of A.

Example. Let $A=\begin{bmatrix} 1 & 2 & -1 \\ 1 & 0 & 1 \\ 4 & -4 & 5 \end{bmatrix}.$

The characteristic polynomial of A is

$$f(\lambda)=\det(\lambda I_n - A)=\begin{vmatrix} \lambda-1 & -2 & 1 \\ -1 & \lambda-0 & -1 \\ -4 & 4 & \lambda-5 \end{vmatrix}$$

$$=-6\lambda^2+11\lambda-6$$

Recall that each term in the expansion of the determinant of an $n \times n$ matrix is a product of n elements of the matrix, containing exactly one element from each row and exactly one element from each column. Thus, if we expand $f(\lambda)=\det(\lambda I_n - A)$, we obtain a polynomial of degree n. A polynomial of degree n with real coefficients has n roots, some of which may be complex numbers. The expression involving λ^n in the characteristic polynomial of A comes from the product

$$(\lambda - a_{11})(\lambda - a_{22})\ldots(\lambda - a_{nn}).$$

So the coefficient of λ^n is 1. We can write

$$f(\lambda)=\det(\lambda I_n - A)=\lambda^n + c_1\lambda^{n-1} + c_2\lambda^{n-2} + \ldots + c_{n-1}\lambda + c_n.$$

If we let $\lambda = 0$ in $\det(\lambda I_n - A)$ as well as in the expression on the right, then we get $\det(-A)=c_n$. Which shows that the constant term c_n is $(-1)^n \det(A)$. This result can be used to establish the following theorem.

Theorem an $n \times n$ matrix A is singular if and only if 0 is an eigenvalue of A.

We now extend our list of Nonsingular Equivalences.

Main points about Nonsingular Equivalences

The following statements are equivalent for an $n \times n$ matrix A.

A is nonsingular.

$Ax = 0$ has only the trivial solution.

A is row equivalent to I_n.

The linear system $Ax = b$ has a unique solution for every $n \times 1$ matrix b.

$\det(A) \neq 0$.

A has rank n.

A has nullity 0.

The rows of A form a linearly independent set of n vectors in R^n.

The columns of A form a linearly independent set of n vectors in R^n.

Zero is not an eigen value of A.

We now connects the characteristic polynomial of a matrix with its eigen values in the following theorem.

Theorem. The eigen values of A are the real roots of the characteristic polynomial of A.

Proof. Let λ be an eigen value of A with associated eigenvector x. Then $Ax = \lambda x,$ which can be rewritten as

$$Ax = (\lambda I_n)x$$

or $$(\lambda I_n - A)x = 0, \qquad (5)$$

a homogeneous system of n equations in n unknowns. This system has a non-trivial solution if and only if the determinant of its coefficient matrix vanishes, that is, if and only if $\det(\lambda I_n - A) = 0.$

Conversely, if λ is a real root of the characteristic polynomial of A, then $\det(\lambda I_n - A) = 0.$ so the homogeneous system (5) has a non-trivial solution x. Hence λ is an eigen value of A.

Thus, to find the eigen values of a given matrix A. we must find the real roots of its characteristic polynomial $f(\lambda).$ There are many methods for finding approximations to the roots of a polynomial, some of them more effective than others; indeed, many computer programs are available to find the roots of a polynomial. Two result that are sometimes useful in this connection are as follows: the product of all the roots of the polynomial

$$f(\lambda) = \lambda^n + a_1\lambda^{n-1} + \ldots + a_{n-1}\lambda + a_n$$

is $(-1)^n \; a_n.$ and if $a_1, a_2, \ldots, a_n$ are integers, then $f(\lambda)$ cannot have a rational root that is not already an integer. Thus, as possible rational roots of $f(\lambda)$, one need only try the integer factor of $a_n.$ Of course, $f(\lambda)$ might well have irrational roots.

However, to minimize the computational effort and as a convenience to the reader, all the characteristic polynomials to be considered in the rest of this chapter have only integer roots, and each of these roots is a factor of the constant term of the characteristic polynomial of A. The corresponding eigen vectors are obtained by substituting the value of λ in equation (5) and solving the resulting homogeneous system. The solution to this type of problem has already discussed.

Example. Consider the matrix $A = \begin{bmatrix} 1 & 2 & -1 \\ 1 & 0 & 1 \\ 4 & -4 & 5 \end{bmatrix}.$

The characteristic polynomial of A is $f(\lambda) = \lambda^3 - 6\lambda^2 + 11\lambda - 6.$

The possible integer roots of $f(\lambda)$ are $\pm 1, \pm 2, \pm 3 \; and \pm 6.$ By substituting these values in $f(\lambda),$ we find that $f(1) = 0,$ so $\lambda = 1$ is a root of $f(\lambda).$ Hence $(\lambda - 1)$ is a factor of $f(\lambda).$ Dividing $f(\lambda)$ by $(\lambda - 1)$, we obtain

$f(\lambda)=(\lambda-1)(\lambda^2-5\lambda+6)$.

Factoring $\lambda^2-5\lambda+6$, we have

$f(\lambda)=(\lambda-1)(\lambda-2)(\lambda-3)$.

The eigen values of A are then

$\lambda_1=1, \lambda_2=2, \lambda_3=3$.

To find an eigen vector x_1 associated with $\lambda_1=1$, we form the system $(1I_3-A)x=0$,

$$\begin{bmatrix} 1-1 & -2 & 1 \\ -1 & 1 & -1 \\ -4 & 4 & 1-5 \end{bmatrix}\begin{bmatrix} x_1 \\ x_2 \\ x_3 \end{bmatrix}=\begin{bmatrix} 0 \\ 0 \\ 0 \end{bmatrix}$$

or $\begin{bmatrix} 0 & -2 & 1 \\ -1 & 1 & -1 \\ -4 & 4 & -4 \end{bmatrix}\begin{bmatrix} x_1 \\ x_2 \\ x_3 \end{bmatrix}=\begin{bmatrix} 0 \\ 0 \\ 0 \end{bmatrix}.$

A solution is $\begin{bmatrix} -\frac{1}{2}r \\ \frac{1}{2}r \\ r \end{bmatrix}$

For any real number r. Thus for $r=2$, $x_1=\begin{bmatrix} -1 \\ 1 \\ 2 \end{bmatrix}$ is an eigen vector of A associated with $\lambda_1=1$. Now to find an eigen vector x_2 associated with $\lambda_2=2$, we form the system

$(2I_3-A)x=0$.

That is, $\begin{bmatrix} 2-1 & -2 & 1 \\ -1 & 2 & -1 \\ -4 & 4 & 2-5 \end{bmatrix}\begin{bmatrix} x_1 \\ x_2 \\ x_3 \end{bmatrix}=\begin{bmatrix} 0 \\ 0 \\ 0 \end{bmatrix}$

A solution is

$$\begin{bmatrix} -\frac{1}{2}r \\ \frac{1}{4}r \\ r \end{bmatrix}$$

For any real number r. Thus, for $r = 4,$

$x_2 = \begin{bmatrix} -2 \\ 1 \\ 4 \end{bmatrix}$ is an eigen vector of A associated with $\lambda_2 = 2.$

Now to find an eigen vector x_3 associated with $\lambda_3 = 3,$ we form the system

$(3I_3 - A)x = 0.$ and find that a solution is $\begin{bmatrix} -\frac{1}{4}r \\ \frac{1}{4}r \\ r \end{bmatrix}$

for any real number r. Thus, for $r = 4,$

$$x_3 = \begin{bmatrix} -1 \\ 1 \\ 4 \end{bmatrix}$$

is an eigen vector of A associated with $\lambda_3 = 3.$

Example. Let $A = \begin{bmatrix} 0 & 1 \\ -1 & 0 \end{bmatrix}.$ Then the characteristic polynomial of A is $f(\lambda) = \lambda^2 + 1,$

Which has no real roots. The roots are $\lambda_1 = i \; and \; \lambda_2 = -i.$ Thus we say that A has no eigen values. Of course, the characteristic polynomial of a matrix may have some complex roots and it may even have no real roots. However, in the important case of symmetric matrices, all the roots of the characteristic polynomial are real.

Eigen values and eigen vectors satisfy many important and interesting properties. For example, if A is an upper(lower) triangular matrix, or a diagonal matrix, then the eigen values of A are the elements on the main diagonal of A. The set S consisting of all eigen vectors of A associated with λ_j as well as the zero vector is a subspace of R^n called the eigen space associated with λ_j.

It must be pointed out that the method for finding the eigen values of a linear transformation or matrix by obtaining the real roots of the characteristics polynomial is not practical for $n > 4$, since it involves evaluating a determinant. Efficient numerical methods for finding eigen values are studied in numerical analysis.

We now turn to examine briefly three applications of eigen values and eigen vectors.

MARKOV Chains

Let T be a regular transition matrix of a Markov process. In a earlier theorem we showed that as $n \to \infty, T^n$ approaches a matrix A, all of whose columns are the identical vector u. Also, showed that u is a steady state vector, which is unique probability vector satisfying the matrix equation $TU = u$. This means that $\lambda = 1$ is an eigen value of T and u is an associated eigen vector. Finally, since the column of A add up to 1, it follows that $\lambda = 1$ is an eigen value of A.

Linear economic models

In Leontief closed model, consisting of a society made up of a farmer, a carpenter, and a tailor, where each person produces one unit of each commodity that is consumed by each individual during the year. The problem facing the economic planner is that of determining the prices $p_1, p_2, and\ p_3$ of the three commodities so that no one makes money or lose money, that is, so that we have a state of equilibrium. Let p denote the price vector. Then the problem is that of finding a solution p to the linear system $Ap = p$ whose components p_1 will be nonnegative with at least one positive p_i. It follows that $\lambda = 1$ is an eigen value of A and p is an associated eigen vector.

Stable Age Distribution in a Population

Consider a population of animals that can live to a maximum age of n years(or any other time unit). Suppose that the number of males in the population is always a fixed percentage of the female population. Thus, in studying the growth of the entire population, we can ignore the male population and concentrate our attention on the female population. We divide the female population into $n+1$ age groups as follows:

$x_i^k =$ number of females of age i who are alive at time $k, 0 \le i \le n$;

$f_i =$ fraction of females of age i who will be alive a year later;

$b_i =$ average number of females born to a female of age i.

Let

$$x^k = \begin{bmatrix} x_0^k \\ x_1^k \\ \vdots \\ x_n^k \end{bmatrix} \qquad (k \ge 0)$$

denote the age distribution vector at time k. The number of females in the first age group at time $k+1$ is merely the total number of females born from time k to time $k+1$. There are x_0^k females in the first age group at time k and each of these females, on the average, produces b_0 female offspring, so the first age group produces a total of $b_0 x_0^k$ females. Similarly, the x_1^k females in the second age group produces a total of $b_1 x_1^k$ females. Thus

$$x_0^{k+1} = b_0 x_0^k + b_1 x_1^k + \ldots + b_n x_n^k \qquad (6)$$

The number x_1^{k+1} of females in the second age group at time $k+1$ is the number of females from the first age group at time k who are alive a year later. Thus

$$x_1^{k+1} = \begin{pmatrix} \textit{fraction of females in} \\ \textit{first age group who are} \\ \textit{alive a year later} \end{pmatrix} \times \begin{pmatrix} \textit{number of females} \\ \textit{in first age} \\ \textit{group} \end{pmatrix},$$

Or $x_1^{k+1} = f_0 x_0^k$,

and, in general

$$x_j^{k+1} = f_{j-1}\ x_{j-1}^k \quad (1 \le j \le n). \tag{7}$$

We can write (6) and (7), using matrix notation as

$$x^{k+1} = Ax^k \quad (k \ge 1) \tag{8}$$

Where

$$A = \begin{bmatrix} b_0 & b_1 & b_2 & \cdots & b_{n-1} & b_n \\ f_0 & 0 & 0 & \cdots & 0 & 0 \\ 0 & f_1 & 0 & \cdots & 0 & 0 \\ \vdots & \vdots & \vdots & \cdots & \vdots & \vdots \\ 0 & 0 & 0 & \cdots & f_{n-1} & 0 \end{bmatrix}$$

and A is called a Leslie matrix. We can use equation (8) to try to determine a distribution of the population by age groups at time $k+1$ so that the number of females in each age group at time $k+1$ will be a fixed multiple of the number in the corresponding age group at time k. That is, if λ is the multiplier, we want $x^{k+1} = \lambda x^k$

or $Ax^k = \lambda x^k$.

Thus λ is an eigen value of A and x^k is a corresponding eigen vector. If $\lambda = 1$, the number of females in each age group will be the same year after year. If we can find an eigen vector x^k corresponding to the eigen value $\lambda = 1$, we say that we have a stable age distribution.

Example. Consider a beetle that can live to a maximum age of two years and whose population dynamics are represented by the Leslie matrix

$$A = \begin{bmatrix} 0 & 0 & 6 \\ \frac{1}{2} & 0 & 0 \\ 0 & \frac{1}{3} & 0 \end{bmatrix}$$

We find that $\lambda = 1$ is an eigen value of A with corresponding eigen vector

$$\begin{bmatrix} 6 \\ 3 \\ 1 \end{bmatrix}$$

Thus, if the numbers of females in the three groups are proportional to $6:3:1$, we have a stable age distribution. That is, if we have 600 females in the first age group, 300 in the second, and 100 in the third, then year after year the number of females in each age group will remain the same.

Population growth problems of the type considered in above example have applications of eigen values and eigen vectors.

Diagonalization

In this section we show how to find the eigen values and associated eigen vectors of a given matrix A by finding the eigen values and eigen vectors of a related matrix B that has the same eigen values and eigen vectors as A. The matrix B has the helpful property that its eigen values are easily obtained. Thus, we will have found the eigen values of A.

Similar Matrices

Definition. A matrix B is said to be similar to a matrix A if there is a nonsingular matrix P such that

$\boldsymbol{B = P^{-1}AP}$.

Example. Let $A = \begin{bmatrix} 1 & 1 \\ -2 & 4 \end{bmatrix}$ be the matrix. Let

$P = \begin{bmatrix} 1 & 1 \\ 1 & 2 \end{bmatrix}$. Then $P^{-1} = \begin{bmatrix} 2 & -1 \\ -1 & 1 \end{bmatrix}$

and

$$B = P^{-1} A P = \begin{bmatrix} 2 & -1 \\ -1 & 1 \end{bmatrix} \begin{bmatrix} 1 & 1 \\ -2 & 4 \end{bmatrix} \begin{bmatrix} 1 & 1 \\ 1 & 2 \end{bmatrix} = \begin{bmatrix} 2 & 0 \\ 0 & 3 \end{bmatrix}.$$

Thus B is similar to A.

Properties for Similarity

Following are the elementary properties for similarity.

A is similar to A.

If B is similar to A, then A is similar to B.

If A is similar to B and B is similar to C, then A is similar to C.

By property 2 we replace the statements "A is similar to B" and " B is similar to A" by " A and B are similar.

Definition. We shall say that the matrix A is diagonalizable if it is similar to a diagonal matrix. In this case we also say that A can be diagonalized.

Example. If A and B are as in above example, then A is diagonalizable, since it is similar to B.

Theorem. Similar matrices have the same eigen values.

Proof. Let A and B be similar. Then $B = P^{-1}AP,$ for some nonsingular matrix P. We prove that A and B have the same characteristic polynomials, $f_A(\lambda)$ and $f_B(\lambda),$ respectively, we have

$$f_B(\lambda) = \det(\lambda I_n - B) = \det(\lambda I_n - P^{-1}AP)$$

$$= \det\left(P^{-1}\lambda I_n P - P^{-1}AP\right) = \det\left(P^{-1}(\lambda I_n - A)P\right)$$

$$= \det\left(P^{-1}\right)\det(\lambda I_n - A)\det(P) \quad (1)$$

$$= \det\left(P^{-1}\right)\det(P)\det(\lambda I_n - A)$$

$$= \det(\lambda I_n - A) = f_A(\lambda).$$

Since $f_A(\lambda) = f_B(\lambda),$ it follows that A and B have the same eigen values.

It suggest that the eigen values of a diagonal matrix are the entries on its main diagonal. The following theorem establishes when a matrix is diagonalizable.

Theorem. An $n \times n$ matrix A is diagonalizable if and only if it has n linearly independent eigen vectors. In this case A is similar to a diagonal matrix D, with $P^{-1}AP = D,$ whose diagonal elements are the eigen values of A, while P is a matrix whose columns are respectively the n linearly independent eigen vectors of A.

Proof. Suppose that A is similar to D. Then

$P^{-1}AP = D.$

So that $AP = PD$ (2)

Let

$$D = \begin{bmatrix} \lambda_1 & 0 & \cdots & 0 \\ 0 & \lambda_2 & \cdots & 0 \\ \vdots & \vdots & \vdots & \vdots \\ 0 & \cdots & 0 & \lambda_n \end{bmatrix},$$

and let $x_j, j = 1, 2, \ldots, n,$ be the jth column of P. Now it follows that jth column of the matrix AP is $Ax_j,$ and the jth column of PD is $\lambda_j x_j.$ Thus from (2) we have

$Ax_j = \lambda_j x_j.$ (3)

Since P is a nonsingular matrix. So by theorem its columns are linearly independent and so are all nonzero. Hence λ_j is an eigen value of A and x_j is a corresponding eigen vector.

Conversely, suppose that $\lambda_1, \lambda_2, \ldots\ldots, \lambda_n$ are n eigenvalue of A and that the corresponding eigenvectors $x_1, x_2, \ldots, x_n$ are linearly independent. Let $p = [x_1, x_2 \ \ , \ldots. \ \ x_n]$ be the matrix whose jth column is x_j column is $x_j.$ Since the columns of p are linearly independent, it follows from that P is nonsingular. From we obtain, which implies that A is diagonalizable. This completes the proof,

Observe that the in the order of the column of P determinant of the order of the columns of P determines the order of the diagonal entries in D.

APPLICATION OF EIGENVALUES AND EIGENVECTORS:

The Fibonacci sequence

In 1202. Leonardo of Pisa, also called Fibonacci, wrote a book on mathematics in which he posed the following problem. A pair of newborn rabbits begins to breed at the age of 1 month. And thereafter produced one pair of offspring per moth suppose that we start with a pair of newly born rabbits and that none of the rabbits produced from this pair die. How many pairs of rabbits will there be at the beginning of each month?

The breeding pattern of the rabbits is shown in. where an arrow indicate an offspring pair. At the begging of month 0, we have the newly born pair of rabbits p_1. At the beginning of month 0, we have the newly born pair of rabbits p_1 which have not yet produced any offspring. At the beginning of month 2. We have the original pair p_2 and its first pair of offspring. p_2. At the beginning of month 3 we have the original pair p_1, its first pair of offspring p_2 born at the beginning of month 2 , and its second pair of offspring. $p3$ At the beggning of month 4 we have P_1, p_2 *and* p_3 ; p_4 , the offspring of p_1; and p_5 the offspring of p_2 . Let u_n denote the number of pairs of rabbits at the beginning of month n. We see that

$$u_0 = 1. \quad u_1 = 1, \quad u_2 = 2, \quad u_3 = 3, \quad u_4 = 5, \quad u_5 = 8.$$

To obtain a formula for u_n. We proceed as follows. The number of pairs of rabbits that are alive at the beginning of month $n\, is\, u_{n-1}$, the number of pairs who were alive at the beginning of month $n\ is u_{n-1}$, the number of pairs who were alive the previous month. Plus the number of pairs newly born at the beginning of month n. The latter number u_{n-2}, *since* a pair of rabbits produces a pair of offspring starting with its second with its month of life. Thus $u_n = u_{n-1} + u_{n-2}$.

That is, each number is the sum of its two Predecessors. The resulting sequence of number, called a Fibonacci sequence occurs in a remarkably varied number of application, such as the distribution of leaves on certain trees, the arrangements of seeds on sunflowers, search technique in numerical analysis generating random numbers in statistics, and other.

To compute u_n by using the recursion relation (or difference equation, we have to compute $u_0, u_1, \ldots\ldots, u_{n-1}$. This can be rather tedious for larger n. We now develop a formula that will enable us to calculate u_n directly. In addition to Equation we write

$$u_{n-1} = u_{n-1},$$

So we now have

$$u_n = u_{n-1} + u_{n-2}$$

$$u_{n-1} = u_{n-1},$$

which can be written in matrix form as

$$\begin{bmatrix} u_n \\ u_{n-1} \end{bmatrix} = \begin{bmatrix} 1 & 1 \\ 1 & 0 \end{bmatrix} \begin{bmatrix} u_{n-1} \\ u_{n-2} \end{bmatrix}.$$

we now define in general

$$w_k = \begin{bmatrix} u_{k+1} \\ u_k \end{bmatrix} \text{ and } A = \begin{bmatrix} 1 & 1 \\ 1 & 0 \end{bmatrix} \quad (0 \le k \le n-1)$$

So that

$$w_0 = \begin{bmatrix} u_1 \\ u_0 \end{bmatrix} = \begin{bmatrix} 1 \\ 1 \end{bmatrix}$$

$$w_1 \begin{bmatrix} u_2 \\ u_1 \end{bmatrix} = \begin{bmatrix} 2 \\ 1 \end{bmatrix}, \ldots\ldots w_{n-1} = A w_{n-2}.$$

Thus

$w_1 = Aw_0$

$w_2 = Aw_1 = A(Aw_0) = A^2\, w_0$

$w_3 = Aw_2 = A(A^2 w_0) = A^3 w_0$

$w_{n-1} = A^{n-1}\, w_0$

Hence, to find u_n, we merely have to calculate A^{n-1}, which rather tedious if n is large. To find avoid the difficulty, we find a diagonal matrix B that is similar to A. The characteristics equation A is

$$\begin{vmatrix} \lambda-1 & -1 \\ -1 & \lambda \end{vmatrix} = \lambda^2 - \lambda - 1 = 0$$

The eigenvalue of A are (verify)

$$\lambda_1 = \frac{1+\sqrt{5}}{2} \quad and\ \lambda_2 = \frac{1-\sqrt{5}}{2}$$

$$D = \begin{bmatrix} \frac{1+\sqrt{5}}{2} & 0 \\ 0 & \frac{1-\sqrt{5}}{2} \end{bmatrix}$$

So

Corresponding of A are (verify)

$$x_1 = \begin{bmatrix} \frac{1-\sqrt{5}}{2} \\ 1 \end{bmatrix} \quad and\ x_2 = \begin{bmatrix} \frac{1-\sqrt{5}}{2} \\ 1 \end{bmatrix}.$$

Thus

$$p = \begin{bmatrix} \frac{1+\sqrt{5}}{2} & \frac{1-\sqrt{5}}{2} \\ 1 & 1 \end{bmatrix}. \qquad p^{-1} = \begin{bmatrix} \frac{1}{\sqrt{5}} & -\frac{1-\sqrt{5}}{2\sqrt{5}} \\ -\frac{1}{\sqrt{5}} & -\frac{1-\sqrt{5}}{2\sqrt{5}} \end{bmatrix}.$$

and $A = PDP^{-1}$

Hence (verify) for any nonnegative integer k,

$A^k = PD^k p^{-1}$.

Since D is diagonal D^k is easy to calculate. Its entries are the diagonal entries of D raised to the kth power. We have

$$w_{n-1} = A^{n-1}w_0 = PD^{n-1}P^{-1}w_0$$

$$=\begin{bmatrix} \frac{1+\sqrt{5}}{2} & \frac{1-\sqrt{5}}{2} \\ 1 & 1 \end{bmatrix}\begin{bmatrix} \left(\frac{1+\sqrt{5}}{2}\right)^{n-1} & 0 \\ 0 & \left(\frac{1-\sqrt{5}}{2}\right)^{n-1} \end{bmatrix}\begin{bmatrix} \frac{1}{\sqrt{5}} & -\frac{1-\sqrt{5}}{2\sqrt{5}} \\ -\frac{1}{\sqrt{5}} & \frac{1+\sqrt{5}}{2\sqrt{5}} \end{bmatrix}\begin{bmatrix} 1 \\ 1 \end{bmatrix}$$

This equation gives the formula (verify)

$$u_n = \frac{1}{\sqrt{5}}\left[\left(\frac{1+\sqrt{5}}{2}\right)^{n-1} - \left(\frac{1-\sqrt{5}}{2}\right)^{n-1}\right]$$

For calculating u_n directly .

Using a handheld calculator, we find that for $n = 50.u_{50}$ us approximately 20.365 billion .2.

DIFFERNTIAL EQUATIONS

A differential equation is an equation that involves an unknown function is

$$\frac{d}{dt}x(t) = rx(t)$$

Where r is a constant. The idea here is to find a function $x(t)$ that will satisfy the given differential equation; that is, $x'(t) = r(t)$. This differential equation is discussed further below.

Differential equations occur often in all branches of science and engineering; linear algebra is helpful in the formulation and survey of the approach; books on differential equations deal with the subject suggestation for further reading are given at the end of this section.

HOMOGENEOUS LINEAR SYSTEMS

We consider the homogeneous linear system of differential equation

$$x_1\ (t) = a_{11}x_1\ (t) + \ a_{12}x_2(t) + \ \ldots + \ a_{1n}x_n(t)$$

$$x_2\ (t) = a_{21}x_1\ (t) + \ a_{22}x_2(t) + \ \ldots + \ a_{2n}x_n(t)$$

$$\vdots$$

$$x_n\ (t) = a_{n1}x_1\ (t) + \ a_{n2}x_2(t) + \ \ldots + \ a_{nn}x_n(t)$$

Where the a_{ij} are known constants. We seek function $x_1(t).x_2(t)\ldots\ldots x_n(t)$ defined and differential on the real line satisfying. We can write in matrix frim by letting

$$x(t) = \begin{bmatrix} x_1(t) \\ x_2(t) \\ \vdots \\ x_n(t) \end{bmatrix}. \qquad A = \begin{bmatrix} a_{11} & a_{12} & \cdots & a_{1n} \\ a_{21} & a_{22} & \cdots & a_{2n} \\ \vdots & \vdots & & \vdots \\ a_{n1} & a_{n2} & \cdots & a_{nn} \end{bmatrix}$$

And defining

$$x(t) = \begin{bmatrix} x_1(t) \\ x_2(t) \\ \vdots \\ x_n(t) \end{bmatrix}$$

Then (1) can be written as

$$x(t) - Ax(t)$$

We shall often write more briefly as

$x = Ax.$

With this notation, an $n-$ vector function

$$x(t) = \begin{bmatrix} x_1(t) \\ x_2(t) \\ \vdots \\ x_n(t) \end{bmatrix}$$

Satisfying is called a solution to the given system. It can be that the set of all solution to the homogeneous linear system of differential equations is a subspace of the vector space of differentiable real-valued $n-vector$ functions.

We leave it to the reader to verify that if $x^1(t).x^2(t)......x^n(t)$ are all solution to then any linear combination

$x(t)=b_1x^1(t)+b_2c^2(t)+......+b_nx^n(t)$ is also a solution to a set of vector functions $Px^1(t)x^2(t)....x^n(t)\}$ is said to be a fundamental system for if every solution to can be written in the form . In this case, the right side of, where $b_1.b_2......b_n$ are arbitrary constants, is said to be the general solution.

It can be shown buy Boyce and $Diprima$ or the book by Cullen cited in Further Readings) that any system of the form has a fundamental system

In general differential equations arise in the course of solving physical problems. Typically. Once a general solution to the differential equation has been obtained the physical constraints in the general solution, giving rise to a particular solution. An important particular solution, is obtained by finding a solution x(t) to such that x(0) – x_0 an initial value problem. If the general solution is known. Then the initial value problem can be solved by setting $t=0$ in and determinating the constants $b_1,b_2,...,b_n$ so that

$$x_0=b_1x^1(0)+b_2x^2(0)+....+b_nx^n(0)$$

It is easily seen that this is actually an $n\times m$ linear system with unknowns $b_1,b_2,...,b_n$. This linear system can also be written as

$$Cb=x_0$$

where

$$b=\begin{bmatrix} b_1 \\ b_2 \\ \vdots \\ b_n \end{bmatrix}$$

\and C is $n \times n$ matrix whose columns are $x^1(0)$, $x^2(0)$, ..., $x^n(0)$, respectively. It can be shown (see the book by Boyce and *Diprima* or the book by Cullen cited in Further Reading) form a fundamental system for, then C is nonsingular, so always has a unique Solution

Example. The simple system of the form (1) is the singular equation $\frac{dx}{dt} = ax,$ Where a is a constant. From calculus, the solutions to this equation are of the form $x = be^{at}$. That is, this is the general solution to. To solve the initial value problem $\frac{dx}{dt} = ax \quad x(0) = x_0$. The system is said to be diagonal if the matrix A is diagonal. Then can be written as

$$x_1^1(t) = a_{11}x_1(t)$$

$$x_2^1(t) = \quad a_{22}x_2(t)$$

$$\vdots$$

$$x_n(t) = \quad a_{nn}x_n(t).$$

This system is easy to solve, since the equations can be solved separately. Applying the results of each equation in. we obtain

$$x_1(t) = b_1 e^{a_{11}t}$$

$$x_2(t) = b_2 e^{a_{22}t}$$

$$\vdots \qquad \vdots$$

$$x_n(t) = b_n e^{a_{nn}t}$$

where $b-1, b_2, \ldots\ldots, b_n$ are arbitrary constant. Writing in vector form yields

$$x(t)=\begin{bmatrix} b_1 e^{a_{11}t} \\ b_2 e^{a_{22}t} \\ \vdots \\ b_n e^{a_{nn}t} \end{bmatrix}=b_1\begin{bmatrix}1\\0\\0\\\vdots\\0\end{bmatrix}e^{a_{11}t},+b_2\begin{bmatrix}0\\1\\0\\\vdots\\0\end{bmatrix}e^{a_{22}t}+\ldots.+b_n\begin{bmatrix}0\\0\\\vdots\\0\\1\end{bmatrix}e^{a_{nn}t}.$$

This implies that the vector functions

$$x^1(t)=\begin{bmatrix}1\\0\\0\\\vdots\\0\end{bmatrix}e^{a_{11}t},\ x^2(t)=\begin{bmatrix}0\\1\\0\\\vdots\\0\end{bmatrix}e^{a_{22}t}\ldots.x^n(t)=\begin{bmatrix}0\\0\\\vdots\\0\\1\end{bmatrix}e^{a_{nn}t}.$$

Form a functional system for the diagonal system

Theorem .If the $n\times n$ matrix A has n linearly independent eigenvectors $P_1,P_2,\ldots\ldots P_n$ associated with the eigenvalue $\lambda_1,\lambda_2\ldots\ldots\lambda_n$. Respectively then the general solution to the homogeneous linear system of difference equations

$x=Ax$ The procedure for obtaining the general solution to the homogeneous linear system $x(t)=Ax(t)$ where A is diagonalizable, is as follows.

Step 1. Compute the eigenvalue $\lambda_1,\lambda_2\ldots..\lambda_n$ *of A.*

Step 2. Compute the eigenvectors $p_1,p_2\ldots\ldots,p_n$ of A associated, respectively, with $\lambda_1,\lambda_2,\ldots.\lambda_n$.

Step 3. The general solution is given by Equation

Example. For the system

$$x=\begin{bmatrix}1 & -1\\ 2 & 4\end{bmatrix}x.$$

The matrix

$$A=\begin{bmatrix}1 & -1\\ 2 & 4\end{bmatrix}$$

Has eigenvalue $\lambda_1 = 2\, and\, \lambda_2 = 3$ with associated eigenvectors(verify)

$$p_1 = \begin{bmatrix} 1 \\ -1 \end{bmatrix} \quad and \quad p_2 = \begin{bmatrix} 1 \\ -2 \end{bmatrix}$$

These eigenvectors are automatically linearly independent. Since they are associated with distinct eigenvalue. Hence the general solution to the given system is

$$x(t) = b_1 \begin{bmatrix} 1 \\ -1 \end{bmatrix} e^{2t} + b_2 \begin{bmatrix} 1 \\ -2 \end{bmatrix} e^{3t}.$$

In terms of components, this can be written as

$$x_1(t) = b_1 e^{2t} + b_2 e^{3t}.$$

$$x_2(t) = -b_1 e^{2t} - 2b_2 e^{3t}.$$

Example. Consider the following homogeneous linear system of differential equations:

$$x = \begin{bmatrix} x_1 \\ x_2 \\ x_3 \end{bmatrix} = \begin{bmatrix} 0 & 1 & 0 \\ 0 & 0 & 1 \\ 8 & -14 & 7 \end{bmatrix} \begin{bmatrix} x_1 \\ x_2 \\ x_3 \end{bmatrix}$$

The Characteristics polynomial of A is (verify)

$$f(\lambda) = \lambda^3 - 7\lambda^2 + 14\lambda - 8$$

Or

$$f(\lambda) = (\lambda - 1)(\lambda - 2)(\lambda - 4).$$

So the eigenvalue of the A are $\lambda_1 = 1, \lambda_2 = 2\ , \lambda_3 = 4.$ Associated eigenvectors are (verify)

$$\begin{bmatrix} 1 \\ 1 \\ 1 \end{bmatrix}, \quad \begin{bmatrix} 1 \\ 2 \\ 4 \end{bmatrix}, \quad \begin{bmatrix} 1 \\ 4 \\ 16 \end{bmatrix}$$

Respectively, the general solution is then given by

$$x(t)=b_1\begin{bmatrix}1\\1\\1\end{bmatrix}e^{t}+b_2\begin{bmatrix}1\\2\\4\end{bmatrix}e^{2t}+b_3\begin{bmatrix}1\\4\\16\end{bmatrix}e^{4t},$$

where $b_1, b_2, and\, b_3$ are arbitrary constants.

Example. For the system of solve the initial value problem determined by the initial conditions $x_1(0)=4, x_2(0)=6,\ and\ x_3(0)=8.$

Solution . We write our general solution in the form $x = p\,\boldsymbol{u}\,as$

$$x(t)=\begin{bmatrix}1&1&1\\1&2&4\\1&4&16\end{bmatrix}\begin{bmatrix}b_1e^{t}\\b_2e^{2t}\\b_3e^{4t}\end{bmatrix}$$

Now

$$x(0)=\begin{bmatrix}4\\6\\8\end{bmatrix}=\begin{bmatrix}1&1&1\\1&2&4\\1&4&16\end{bmatrix}\begin{bmatrix}b_1e^{0}\\b_2e^{0}\\b_3e^{0}\end{bmatrix}$$

or

$$\begin{bmatrix}1&1&1\\1&2&4\\1&4&16\end{bmatrix}\begin{bmatrix}b_1\\b_2\\b_3\end{bmatrix}=\begin{bmatrix}4\\6\\8\end{bmatrix}.$$

Solving by Gauss-Jordan reduction, we obtain (verify)

$$b-1=\frac{4}{3},\quad b_2=3\ ,\quad b_3=-\frac{1}{3}$$

Therefore, the solution to the initial value problem is

$$x(t)=\frac{4}{3}\begin{bmatrix}1\\1\\1\end{bmatrix}e^{t}+3\begin{bmatrix}1\\2\\4\end{bmatrix}e^{2t}-\frac{1}{3}\begin{bmatrix}1\\4\\16\end{bmatrix}e^{4t}.$$

We now recall several facts from. If A does not have distinct eigenvaslues, then we may not be able to diagonalized if and only if the dimension of the eigen space associated with $\lambda\, is\, k, that$ is , if and only if the rank of the matrix $(\lambda I_n - A)\, is\, n-k$ (verify). If the rank of $(\lambda I_n - A)\, is\, n-k$. Then we can find k linealry independent eigenvectors of A associated with λ.

Example. Consider the linear system

$$x' = Ax = \begin{bmatrix} 1 & 0 & 0 \\ 0 & 3 & -2 \\ 0 & -2 & 3 \end{bmatrix} x.$$

The eigenvalue of A are $\lambda_1 = \lambda_2 = 1\, and\, \lambda_3 = 5$. The rank of the matrix

$$(1I_3 - A) = \begin{bmatrix} 0 & 0 & 0 \\ 0 & -2 & 2 \\ 0 & 2 & 2 \end{bmatrix}$$ is 1

and the linearly independent eigenvectors $\begin{bmatrix} 1 \\ 0 \\ 0 \end{bmatrix} and \begin{bmatrix} 0 \\ 1 \\ 1 \end{bmatrix}$

Are associated with the eigenvalue 1. The eigenvector $\begin{bmatrix} 0 \\ 1 \\ -1 \end{bmatrix}$ is associated with the eigenvalue 5. The general solution to the given system is then

$$x(t0= b_1 \begin{bmatrix} 1 \\ 0 \\ 0 \end{bmatrix} e' + b_2 \begin{bmatrix} 0 \\ 1 \\ -1 \end{bmatrix} e' + b_3 \begin{bmatrix} 0 \\ 1 \\ -1 \end{bmatrix} e^{5t}$$

Where $b_1, b_2,\, and\, b_3$ are arbitrary constants

If we diagonalized A as in the example we have a considerably more difficult situation. Methods for dealing with such problems are discussed in more advanced books.

SUMMARY

In this chapter we discussed about Eigen values, eigen vectors and their application in real life. A question of considerable importance in a great many applied problems is the determination of vector x, if there are any, such that $x \, and \, Ax$ are parallel. Such questions arise in all applications involving vibrations; chemical engineering, biology, differential equations and so on. In this section we shall formulate this problem precisely; we also define some pertinent terminology. In the next section we solve this problem for symmetric matrices and briefly discuss the situation in the general case.

Definition. Let $A be an n \times n$ matrix. The real number λ is called an eigenvalue of A if there exists a nonzero vector x in R^n such that

$$Ax = \lambda x. \qquad (1)$$

Every non zero vector x satisfying (1) is called an eigenvector of A associated with the eigenvalue λ. We might mention that the word "eigenvalue" is a hybrid one ("eigen" in German means "proper"). Eigen values are also called proper values, characteristic values, and latent values; and eigen vectors are also called proper vectors, and so on, accordingly.

Theorem. An $n \times n$ matrix A is singular if and only if 0 is an eigenvalue of A.

Main points about Nonsingular Equivalences

The following statements are equivalent for an $n \times n$ matrix A.

A is nonsingular.

$Ax = 0$ has only the trivial solution.

A is row equivalent to I_n.

The linear system $Ax = b$ has a unique solution for every $n \times 1$ matrix b.

$\det(A) \neq 0$.

A has rank n.

A has nullity 0.

The rows of A form a linearly independent set of n vectors in R^n.

The columns of A form a linearly independent set of n vectors in R^n.

Zero is not an eigen value of A.

Theorem. The eigen values of A are the real roots of the characteristic polynomial of A.

Three important and real life problem application of eigen values and eigen vectors.

Markov chains

Linear economic models

Stable Age Distribution in a Population

Similar Matrices.

Definition. A matrix B is said to be similar to a matrix A if there is a nonsingular matrix P such that $\boldsymbol{B} = \boldsymbol{P}^{\ 1}\boldsymbol{AP}$.

Properties for Similarity

Following are the elementary properties for similarity.

A is similar to A.

If B is similar to A, then A is similar to B.

If A is similar to B and B is similar to C, then A is similar to C.

Definition. We shall say that the matrix A is diagonalizable if it is similar to a diagonal matrix. In this case we also say that A can be diagonalized.

Theorem. Similar matrices have the same eigen values.

Theorem. An $n \times n$ matrix A is diagonalizable if and only if it has n linearly independent eigen vectors. In this case A is similar to a diagonal matrix D, with $P^{-1}AP = D$, whose diagonal elements are the eigen values of A, while P is a matrix whose columns are respectively the n linearly independent eigen vectors of A.

Application of Eigen Values and Eigen Vectors

The Fibonacci Sequence $u_n = u_{n-1} + u_{n-2}$. That is, each number is the sum of its two Predecessors. The resulting sequence of number, called a Fibonacci sequence occurs in a remarkably varied number of application, such as the distribution of leaves on certain trees, the arrangements of seeds on sunflowers, search technique in numerical analysis generating random numbers in statistics, and other.

To compute u_n by using the recursion relation (or difference equation, we have to compute $u_0, u_1, \ldots, u_{n-1}$. This can be rather tedious for larger n. We now develop a formula that will enable us to calculate u_n directly.

Differntial Equations

A differential equation is an equation that involves an unknown function is

$$\frac{d}{dt}x(t) = rx(t)$$

Where r is a constant. The idea here is to find a function $x(t)$ that will satisfy the given differential equation; that is, $x'(t) = r(t)$. This differential equation is discussed further below.

Differential equations occur often in all branches of science and engineering; linear algebra is helpful in the formulation and survey of the approach

A set of vector functions $Px^1(t)x^2(t)\ldots x^n(t)\}$ is said to be a fundamental system for if every solution to can be written in the form. In this case, the right side of, where $b_1.b_2\ldots\ldots b_n$ are arbitrary constants, is said to be the general solution.

In general. Differential equations arise in the course of solving physical problems. Typically. Once a general solution to the differential equation has been obtained the physical constraints in the general solution, giving rise to a particular solution.

Theorem. If the nxn matrix A has n linearly independent eigenvectors $P_1, P_2, \ldots\ldots P_n$ associated with the eigenvalue $\lambda_1, \lambda_2 \ldots\ldots \lambda_n$. Respectively then the general solution to the homogeneous linear system of difference equations

$x = Ax$ The procedure for obtaining the general solution to the homogeneous linear system $x(t) = Ax(t)$ where A is diagonalizable, is as follows.

Step 1. Compute the eigenvalue $\lambda_1, \lambda_2 \ldots\ldots \lambda_n$ *of A*.

Step 2. Compute the eigenvectors $p_1, p_2 \ldots\ldots, p_n$ of A associated, respectively, with $\lambda_1, \lambda_2, \ldots\ldots \lambda_n$.

Step 3. The general solution is given by Equation.

CHAPTER 7

SOLUTION OF LINEAR SYSTEM OF EQUATIONS AND MATRIX INVERSION

INTRODUCTION

Many real-life problems in engineering give rise to a system of linear equations. *For example*, such systems occurs in certain applications of statistical analysis and in finding the numerical solution of partial differential equations and so on, it is therefore natural to seek efficient methods for solving these equations numerically .

The generally from of a system of m linear equations in n unknowns $x_1, x_2, x_3 \ldots,$ can be represented in matrix from as under:

$$\begin{bmatrix} a_{11}\, a_{12}\, a_{13 \ldots\ldots\ldots} a_{1n} \\ a_{21}\, a_{22}\, a_{23 \ldots\ldots} a_{2n} \\ . \\ . \\ . \\ a_{m1}\, a_{m2}\, a_{m3} \ldots a_{mn} \\ . \end{bmatrix} \begin{pmatrix} x_1 \\ x_2 \\ . \\ . \\ . \\ x_n \end{pmatrix} = \begin{pmatrix} b_1 \\ b_2 \\ . \\ . \\ . \\ b_m \end{pmatrix}$$

Using matrix notation, the above system can be written in compact form as

$$[A](X) = (B)$$

The solution of the system of equation gives n unknown values $x_1, x_2 \ldots, x_m$, which satisfy the system simultaneously, if $m > n$, we may not be able to find a solution, in principle, which satisfy all the equations. If $m > n$, the system usually will have an infinite number of solutions. However, in this chapter, we shall restrict to the case $m = n$. In this case, $if\,|A| \neq$, then the system will have a unique solution, while $|A| = 0$, then there exist no solution.

Various numerical methods are available for finding the solution of the system of equations and they are classified as direct and iterative methods. In direct methods, we get the solution of the system after performing all the steps involved in the procedure. The direct methods consist of elimination methods and decomposition methods. In this chapter, under elimination methods, we consider, Gaussian elimination and Gauss-Jordan elimination methods. Crout's reduction also known as Cholesky's reduction is considered under decomposition methods. Under iterative methods, the initial approximate solution is assumed to be known and is improved towards the exact solution in an iterative methods. All these methods are easily adoptable to computers and can be used to solve even hundred or more simultaneous linear equations.

GAUSSIAN ELIMINATION METHOD

In the Gaussian elimination method, the solution to the system of Equations is obtained in two stages. In the first stage, the given system of equations is reduced to an equivalent upper triangle system is solved using back substitution procedure by which we obtain the solution in the order $x_n, x_{n-1}, x_{n-2}, \ldots\ldots x_2, x_1$

This method is explained by considering a system of n equations in n unknowns in the form as follows.

$$\left.\begin{array}{l} a_1 x_1 + a_{12} x_2 + \ldots + a_{1n} x_n = b_1 \\ a_{21 x_1} + a_{22} x_2 + \ldots.. a_{2n} x_n = b_2 \\ \qquad \ldots \\ \qquad \ldots \\ \qquad \ldots \\ a_n x_1 + a_{n2} x_2 + \ldots.. a_{nn} x_n = b_n \end{array}\right\}$$

Stage1: We divide the first equation by a_{11} and then subtract this equation multiplied by $a_{21}, a_{31}, \ldots, a_{n1}$ form the 2nd, 3rd, ..., $n-th$ the equation . Then the

$$\left.\begin{array}{l} x_1 + a_{12}x_2 + \ldots + a_{1n}x_n = b_1 \\ \quad a_{22}x_2 + \ldots .. a_{2n}x_n = b_2 \\ \quad\quad \ldots \\ \quad\quad \ldots \\ \quad\quad \ldots \\ \quad a_{n2}x_2 + \ldots .. a_{nn}x_n = b_n \end{array}\right\}$$

Here, we can observe that the last $(n-1)$ equations are independent of x_1, that is, x_1 is eliminated from the last $(n-1)$ equations,

This procedure is repeated with the second equation of that is, we divide the second equation by a_{22} and the x_2 is eliminated from 3rd, 4th, ..., $n-th$ equations of the same procedure is repeated again and again till the given system assumes the following upper triangular form:

$$\left.\begin{array}{l} c_{11}x_1 + c_{12}x_2 + \ldots + c_{1n}x_n = d_1 \\ \quad c_{22}x_2 + \ldots .. c_{2n}x_n = d_2 \\ \quad\quad \ldots \\ \quad\quad \ldots \\ \quad\quad \ldots \\ \quad\quad c_{nn}x_n = d_n \end{array}\right\}$$

Stage II: Now, the values of the unknowns are determined by back substitution procedure, in which we obtain x_n from the last equation of and then substituting this value of x_{n-1}. Continuing this way, we can find the values of all other unknowns in the order $x_n, x_{n-1}, \ldots x_2, x_1$. In this method, we observe that the determine of the coefficient matrix is obtained as a by-product, that is $|A| c_{11}c_{22} \ldots c_{nn}$. To familiarize with the method, we consider the following example:

Example 1

Solve the following system of equations using Gaussian elimination method

$$2x+3y-z=5$$

$$4x+4y-3z=3$$

$$-2x+3y-z=1$$

Solution

The given system of equation is solved in two stages

Stage 1 (reduction to upper-triangular from): We divide the first equation by 2 and then subtract the resulting equation (multiplied by 4 and -2) form the second and third equations respectively. Thus, we eliminate x from the 2nd and 3rd equations. The resulting new system given by

$$\left.\begin{aligned} x+\frac{3}{2}y-\frac{z}{2}=\frac{5}{2} \\ -2y-z=-7 \\ 6y-2z=6 \end{aligned}\right\}$$

Now, we divide the second equation of (1) by -2 and eliminate from the last equation and the modified system is given by

$$\left.\begin{aligned} x+\frac{3}{2}y-\frac{z}{2}=\frac{5}{2} \\ y+\frac{z}{2}=\frac{7}{2} \\ -5z=-15 \end{aligned}\right\}$$

Stage II (back substitution): from the last equation of (2), we immediately get

$$Z=3$$

Using this value of z, the second equation of (2) gives

$$y=\frac{7}{2}-\frac{3}{2}=2$$

Using this value of y and z in the first equation of (2), we get

$$x = \frac{5}{2} + \frac{3}{2} - 3 = 1$$

Thus, the solution of the given system is given by Eqs

Partial and full pivoting

The Gaussian elimination method fails if any one of the pivot elements becomes zero. In such a situation, we rewrite the equations in a different order to avoid zero pivots. Changing the order of equations is called *pivoting.*

We now introduce the concept of partial pivoting. In this technique, if the pivot a_{ii} happens to be zero, then the i-th column elements are searched for the numerically largest elements. Let the j-th row (j> i) contains this element, then we interchange the i-th equation with the j-th equation and proceed for elimination. For example, let us examine the solution of the following simple system

$$10^{-5} x_1 + x_2 = 2$$

$$x_1 + x_2 = 2$$

Using Gaussian elimination method with and without partial pivoting, assuming that we require the solution accurate to only four decimal places. The solution by Gaussian elimination gives $x_1 = 0, x_2 = 1.$ if we use partial pivoting, the system takes the form

$$x_1 + x_2 = 2$$

$$10^{-5} x_1 + x_2 = 1$$

Using Gaussian elimination method, the solution is found to $x_1 = 1, x_2 = 1,$ which is meaning full and perfect result.

In full pivoting which is also known as complete pivoting, we interchange rows as columns, such that the largest element in the matrix of the system becomes the pivot element. In this process, the position of the unknown variable also get changed. Full pivoting, in fact, is more complicated that the partial pivoting. Partial pivoting is preferred for hand computation.

Example 2

Solve the system of equations

$$x + y + z = 7$$

$$3x + 3y + 4z = 24$$

$$2x + y + 3z = 16$$

by Gaussian elimination method with partial pivoting.

Solution. In matrix notation, the given system can be written as

$$\begin{bmatrix} 111 \\ 334 \\ 213 \end{bmatrix} \begin{pmatrix} x \\ y \\ z \end{pmatrix} = \begin{pmatrix} 7 \\ 24 \\ 16 \end{pmatrix}$$

To start with, we observe that the pivot element $a_{11} = 1(\neq 0)$. However, a glance at the first column reveals that the numerically largest element is 3 which is in the second row. Hence, we interchange the first row with the second row and then proceed for elimination. Thus, Eq take the form

$$\begin{bmatrix} 334 \\ 111 \\ 213 \end{bmatrix} \begin{pmatrix} x \\ y \\ z \end{pmatrix} = \begin{pmatrix} 24 \\ 7 \\ 16 \end{pmatrix}$$

After partial pivoting.

Stage1(reduction to upper triangular from): By dividing the first row of the system and then subtracting the resulting row, multiplied by 1 and 2 from the second and third rows of the system (2), we get

$$\begin{bmatrix} 11\frac{4}{3} \\ -\frac{1}{3} \\ -1\frac{1}{3} \end{bmatrix} \begin{pmatrix} x \\ y \\ z \end{pmatrix} = \begin{pmatrix} 8 \\ -1 \\ 0 \end{pmatrix}$$

The second row in $Eq.$ cannot be used as the pivot row, as $a_{22} = 0.$ Interchanging the second and third rows, we obtain

$$\begin{bmatrix} 11\frac{4}{3} \\ -1\frac{1}{3} \\ -\frac{1}{3} \end{bmatrix} \begin{pmatrix} x \\ y \\ z \end{pmatrix} = \begin{pmatrix} 8 \\ 0 \\ -1 \end{pmatrix}$$

Which is in the upper triangular from.

Stage II: From the last row of Eq we at once get

$z = 3$

The second row of $Eq.$ with this value of z gives

$y+1=0 \, or \, y=1$

Using these values of y and z, the first row of $Eq.$ gives

$x+1+4=8 \, or \, x=3$

Thus Eqs . Constitute the solution to the given system of equations.

Example 3.

Solve by Gaussian elimination method with partial pivoting, the following system of equations:

$0x_1 + 4x^2 + 2x_3 + 8x_4 = 24$

$4x_1 + 10x_2 + 5x_3 + 4x_4 = 32$

$4x_1 + 5x_2 + 6.5x_3 + 2x_4 = 26$

$9x_1 + 4x_2 + 4x_3 +)x_4 = 21$

Solution

In matrix notation, the given system can be written as

$$\begin{bmatrix} 04\ 2\ 8 \\ 41054 \\ 456.52 \\ 94\ 40 \end{bmatrix} \begin{pmatrix} x_1 \\ x_2 \\ x_3 \\ x_4 \end{pmatrix} = \begin{pmatrix} 21 \\ 32 \\ 26 \\ 21 \end{pmatrix}$$

To start with, we observe that the pivot row, that is, the first row has a zero pivot element $(a_{11} = 0).$ this row should be interchanged with any row following it, which on becoming a pivot row should not have a zero pivot element. While interchanging rows it is better to interchange with a row having largest pivotal element. Thus, we interchange the first and fourth row, which is called partial pivoting and get,

$$\begin{bmatrix} 94\ 4\ 0 \\ 41054 \\ 456.52 \\ 04\ 28 \end{bmatrix} \begin{pmatrix} x_1 \\ x_2 \\ x_3 \\ x_4 \end{pmatrix} = \begin{pmatrix} 21 \\ 32 \\ 26 \\ 24 \end{pmatrix}$$

We observe that, in partial pivoting, the unknowns' vector remains unaltered, while the right-hand side vector gets changed

Now, we shall carry out Gaussian elimination process in two stages.

Stage 1

(reduction to upper-triangular from): In this stage, by dividing the first row of the system (2) by 9 and then subtracting this resulting row, multiplied by 4 and 4 from the second and third rows of Eq we get

$$\begin{bmatrix} 1\ \frac{4}{9}\ \frac{4}{9}\ 0 \\ 08.22223.22224 \\ 03.22224.72222 \\ 0\ 4\ 4\ 0 \end{bmatrix} \begin{pmatrix} x_1 \\ x_2 \\ x_3 \\ x_4 \end{pmatrix} = \begin{pmatrix} 2.3333 \\ 22.6666 \\ 16.6666 \\ 24 \end{pmatrix}$$

Now we divide the second pivot row by 8.2222 and subtract the resultant row multiplied by 3.2222 and 4 from the third and fourth rows of Eq. to get

$$\begin{bmatrix} 9 & \frac{4}{9} & \frac{4}{9} & 0 \\ 4 & 1 & 3.2222 & 0.4865 \\ 4 & 0 & 3.4594 & 0.4324 \\ 0 & 0 & 0.4324 & 6.0540 \end{bmatrix} \begin{pmatrix} x_1 \\ x_2 \\ x_3 \\ x_4 \end{pmatrix} = \begin{pmatrix} 2.3333 \\ 2.7568 \\ 7.7836 \\ 12.9728 \end{pmatrix}$$

Finally, we divide the third pivot row by 3.4594 and subtract the resultant row multiplied by 0.4324 from fourth row of Eq thus getting the upper triangular from

$$\begin{bmatrix} 9 & \frac{4}{9} & \frac{4}{9} & 0 \\ 4 & 1 & 3.2222 & 0.4865 \\ 4 & 0 & 1 & 0.1250 \\ 0 & 0 & 0 & 5.9999 \end{bmatrix} \begin{pmatrix} x_1 \\ x_2 \\ x_3 \\ x_4 \end{pmatrix} = \begin{pmatrix} 2.3333 \\ . \\ 2.7568 \\ 2.2500 \\ 11.9999 \end{pmatrix}$$

Stage II. (Black Substitution): From the last row of $Eq.$ we immediately get $x_4 = 2.0000.$ Using this value of x_4 *into* the third row of $Eq.$ we obtain

$x_3 + 0.25 = 2.25 \, or \, x_3 = 2.0000$

Similarly, we get

$x_2 = 1.0000, x_1 = 1.0000$

Thus, the solution of the given system is given by

$x_1 = 1.0,\ x_2 = 1.0.\ x_3 = 2.0,\ x_4 = 2.0$

GAUSS-JORDAN ELIMINATION METHOD

This method is a variation of Gaussian elimination method, In this method, the elements above and below the diagonal are simultaneously made zero and thereby the given system is reduced to a equivalent diagonal from using elementary transformations. Then the solution of the resulting diagonals system an be readily obtained.

Sometimes, we normalize the pivot row with respect to the pivot element, before elimination. Partial pivoting is also used whenever the pivot element becomes zero.

This method is illustrated through the following examples

Example 4 Solve the system of equations

$$\left.\begin{aligned} x+2y+z&=8 \\ 2x+3y+4z&=20 \\ 4x+3y+2z&=16 \end{aligned}\right\}$$

Using Gauss-Jordan elimination method.

Solution

In matrix notation, the given system (1) can be written as

$$\begin{bmatrix} 121 \\ 234 \\ 432 \end{bmatrix}\begin{pmatrix} x \\ y \\ z \end{pmatrix}=\begin{pmatrix} 8 \\ 20 \\ 16 \end{pmatrix}$$

We subtract the first row multiplied by 2 and 4 from the second and third rows respectively of $Eq.$ and eliminate x

$$\begin{bmatrix} 1\,2\,1 \\ 0\,-1\;2 \\ 0\,-5-2 \end{bmatrix}\begin{pmatrix} x \\ y \\ z \end{pmatrix}\begin{pmatrix} 8 \\ 4 \\ -16 \end{pmatrix}$$

Now , we eliminate y from the first and third rows using the second row. Thus, we get

$$\begin{bmatrix} 1\,0\,1 \\ 0\,-1\,2 \\ 05-12 \end{bmatrix}\begin{pmatrix} x \\ y \\ z \end{pmatrix}\begin{pmatrix} 16 \\ 4 \\ -36 \end{pmatrix}$$

Before, eliminating z from the first and second row, normalizing the third row with respect to the pivot element, we get

$$\begin{bmatrix} 1\,0\,5 \\ 0\,-1\,2 \\ 0\,0\,1 \end{bmatrix}\begin{pmatrix} x \\ y \\ z \end{pmatrix}\begin{pmatrix} 16 \\ 4 \\ 3 \end{pmatrix}$$

Using, the third row of $Eq.$ eliminating z from the first and second rows of $Eq.$

$$\begin{bmatrix} 1 & 0 & 0 \\ 0 & -1 & 0 \\ 0 & 0 & 1 \end{bmatrix} \begin{pmatrix} x \\ y \\ z \end{pmatrix} \begin{pmatrix} 1 \\ -2 \\ 3 \end{pmatrix}$$

From $Eq.$, we get the solution directly as $x = 1, y = 2, z = 3.$

CROUT'S REDUCTION METHOD

The method is based on the fact that the coefficient matrix $[A]$ of the system of equations can be decomposed into the product of two matrices [L] and [U], Where [L] is a lower-triangular matrix and [U] is an upper triangular matrix with 1's on its main diagonal. The rules for getting [L] and [U] can be obtained from the fact

$$[L][U]=[A]$$

For the purpose of illustration, let us consider a (3×3) general matrix in the form

$$\begin{bmatrix} l_{11} & 0 & 0 \\ l_{12} & l_{22} & 0 \\ l_{31} & l_{32} & l_{33} \end{bmatrix} \begin{bmatrix} 1\frac{11}{12}\frac{11}{13} \\ 01\frac{11}{23} \\ 001 \end{bmatrix} = \begin{bmatrix} a_{11} & a_{12} & a_{13} \\ a_{21} & a_{22} & a_{23} \\ a_{31} & a_{32} & a_{33} \end{bmatrix}$$

The sequence of steps for getting [L] and [U] are given below:

Step 1: Multiplying all the rows of [L] is same as the first column of [U], we get

$$l_{11} = a_{11}, l_{21} = a_{21}, l_{31} = a_{31}$$

Thus, we observe that the first column of [L] is same as the first column of [A].

Step II. Now, multiplying the first row of [L] by the second column of [U] , we obtain

$$l_{11}u_{12} = a_{12}\ ,\quad l_{11}u_{13} = a_{13}$$

or

$$ll_{12} = \frac{a_{12}}{l_{11}}\ ,\ u_{13} = \frac{a_{13}}{l_{11}}$$

Thus, the first row of [U] is obtained Now, we continue this process getting alternately the column of [L] and a row of [U] .

Step III. Multiply the second and third rows of [L] by the second column [U] to get

$$l_{21}u_{12} + l_{22n} = a_{22},\ l_{31}u_{12} + l_{32} = a_{32}$$

Which gives

$$l_{22} = a_{22} - l_{21}\,u_{12},\quad l_{32} = a_{32} - l_{31}u_{12}$$

Thus, second column of [L] is obtained.

Step Iv: Now, multiply the second row of [L] by the third column of [U] which yields

$$L_{12}u_{13} + l_{32}u_{23} = a_{23} \quad or\ u_{23} = \frac{a_{23} - l_{21}u_{13}}{l_{22}}$$

Step V:

Lastly, we multiply the third row of [L] by the third columns of [U] and get

$$l_{31}\,u_{13} + l_{32}u_{23} + l_{33} = a_{33}$$

Which gives

$$l_{33} = a_{33} - l_{31}u_{13} - l_{32}u_{23}$$

Thus, the above five steps determine [L] and [U]. This algorithm can generalized to any linear system of order n.

Now, to obtain the solution of the linear system

$$\left.\begin{aligned} a_{11}x_1 + a_{12}x_2 + a_{13}x_3 = b_1 \\ a_{21}x_1 + a_{22}x_2 + a_{23}x_3 = b_2 \\ a_{31}x_1 + a_{32}x_2 + a_{33}x_3 = b_3 \end{aligned}\right\}$$

In matrix notation as $[A](X)=(B)$. Let $[A]=[L][U]$, then we get

$[L][U](X)=(B)$

Substituting $[U](X)=(Z)\, in\, Eq$ we obtain

$[L](Z)=(B)$

Now, Eq is equivalent to

$$\left.\begin{array}{r} l_{11}z_1 = b_1 \\ l_{21}z_1 + l_{22}z_2 = b_2 \\ l_{31}z_1 + l_{32}z_2 + l_{32}z_3 = b_3 \end{array}\right\}$$

The first of these equation gives z_1. Knowing z_1, the second equation of given $z_2; then\, the\, third\, equation\, of$ can be compute $x_1, x_2\, and\, x_3$ from equation [U] (x) = (z) or from

$$\begin{bmatrix} 1\; l_{12}\; u_{13} \\ 0\; 1\; u_{23} \\ 0\; 0\; 1 \end{bmatrix} \begin{pmatrix} x_1 \\ x_2 \\ x_3 \end{pmatrix} = \begin{pmatrix} z_1 \\ z_2 \\ z_3 \end{pmatrix}$$

This method is also known as Cholesky's reduction method. This technique is widely used in the numerical solution of partial differential equations.

This method is very popular from computer programming point of view, since the storage space reserved for matrix [A] can be used to store the elements of [L] and [U] at the end of computation

It may be noted that this method fails if any $a_{ii} = 0.$ in this case, the system in singular. In order to familiarize with the Crout's reduction method, we consider the following examples

Example 5

Solve the following system of equations

$5x_1 - 2x_2 + x_3 = 4$

$7x_1 + x_2 - 5x_3 = 8$

$3x_1 + 7x_2 + 4x_3 = 10$

By Crout's reduction method using hand computation.

Solution Let the coefficient matrix [A] be written as [L] [U]. Thus,

$$\begin{bmatrix} l_{11} & 0 & 0 \\ l_{21} & l_{22} & 0 \\ l_{31} & L_{32} & L_{33} \end{bmatrix} \begin{bmatrix} 1 & u_{12} & u_{13} \\ 0 & 1 & u_{23} \\ 0 & 0 & 1 \end{bmatrix} = \begin{bmatrix} 5 & -2 & 1 \\ 7 & 1 & -5 \\ 3 & 7 & 4 \end{bmatrix}$$

Step 1:

Multiply all the rows of [L] by the first column of [U], we get

$l_{11} = 5, l_{21} = 7, l_{31} = 3$

Step II:

Multiply the first row of [L] by the second and third columns of [U], we have

$l_{11}u_{12} = -2$, $l_{11}u_{13} = 1$

Using Eq. we get

$$u_{12} = -\frac{2}{5}, u_{13} = \frac{1}{5}$$

Step III: Multiply the second and third rows of [L] by the second column of [U]. Using $Eqs.$ We get

$$\left.\begin{aligned} l_{21}u_{12} + l_{22} &= 1 \; or \; l_{22} = 1 + \frac{14}{5} = \frac{19}{5} \\ l_{31}u_{12} + l_{32} &= 7 \; or \; l_{32} = 7 + \frac{6}{5} = \frac{41}{5} \end{aligned}\right\}$$

Step IV: Multiply the second row of [L] by the third column of [U] which yields $l_{21}u_{13} + l_{22}i_{23}$ = - 5. Using $Eqs.$, we get

$$\frac{19}{5}u_{23} = -5 - \frac{7}{5}$$

Therefore,

$$u_{23} = -\frac{32}{19}$$

Step V:

Finally, multiply the third row of [L] with the third column of [U], we obtain

$$l_{31}u_{13} + l_{32}\,u_{23} + l_{33} = 4$$

Using $Eqs. we\, get$

$$l_{32} = \frac{327}{19}$$

Thus, the given system of equations takes the form $[L][U][X]=(B)$. That is,

$$\begin{bmatrix} 500 \\ 7\frac{19}{5}0 \\ 3\frac{41}{5}\frac{327}{19} \end{bmatrix} \begin{bmatrix} 1-\frac{2}{5}\frac{1}{5} \\ 0\,1-\frac{32}{19} \\ 00\,1 \end{bmatrix} \begin{pmatrix} x_1 \\ x_2 \\ x_3 \end{pmatrix} = \begin{pmatrix} 4 \\ 8 \\ 10 \end{pmatrix}$$

Let [U] (x) = (z), them

$$[L](Z) = (4810)^T$$

or

$$\begin{bmatrix} 500 \\ 7\frac{19}{5}0 \\ 3\frac{41}{5}\frac{327}{19} \end{bmatrix} \begin{pmatrix} z_1 \\ z_2 \\ z_3 \end{pmatrix} = \begin{pmatrix} 4 \\ 8 \\ 10 \end{pmatrix}$$

Which gives

$$z_1 = \frac{4}{5},\ z_2 = \frac{12}{19}, z_3 = \frac{46}{327}$$

Utilizing these values of z, ^{Eq} becomes

$$\begin{bmatrix} 1-\frac{2}{5}\frac{1}{5} \\ 01-\frac{32}{19} \\ 001 \end{bmatrix} \begin{pmatrix} x_1 \\ x_2 \\ x_3 \end{pmatrix} = \begin{pmatrix} \frac{4}{5} \\ \frac{12}{19} \\ \frac{46}{327} \end{pmatrix}$$

By back substitution method, we obtain

$$x_3 = \frac{46}{327}, x_2 = \frac{284}{327}, x_1 = \frac{366}{327}$$

This is the required solution.

JACOBI'S METHOD

Jacobi's method is an iterative method, where initial approximate solution to a given system of equations is assumed and is improved towards the exact solution in an iterative way. In general, when the coefficient matrix of the system of equations is a sparse matrix (many elements are zero), iterative methods have definite advantage over direct methods in respect of economy in computer memory. Such spare matrices arise in computing the numerical solution of partial differential equations.

To illustrate Jacobi's method, let us consider a linear system given by

$$\left.\begin{array}{l} a_{11}x_1 + a_{12}x_2 + \ldots. + a_{1n}\, x_n = b_1 \\ \quad\quad\quad . \\ a_{21}x_1 + a_{22}x_2 + \ldots. + 1_{2n}\, x_n = b_2 \\ \quad\quad\quad . \\ a_{n1}x_1 + a_{n2}x_2 + \ldots. + a_{nn}\, x_n = b_n \end{array}\right\}$$

In this method, we assume that the coefficient matrix [A] the modulus of the diagonal element exceeds the sum of the off-diagonal elements. We also assume that the diagonal element vanishes, the equations can always be rearranged to satisfy this condition. Now the system can be written as

$$\left.\begin{aligned} x_1 &= \frac{b_1}{a_1} x_2 - \ldots - \frac{a_{in}}{a_{11}} x_n \\ x &= \frac{b_2}{a_{22}} - \frac{a_{21}}{a_{22}} x_1 - \ldots - \frac{a_{2n}}{a_{22}} x_n \\ x_n &= \frac{b_n}{a_{nn}} - \frac{a_{n1}}{a_{nn}} x_1 - \ldots - \frac{a_n(n-1)}{a_{nn}} x_{n-1} \end{aligned}\right\}$$

We shall take this solution vector $(x_1, x_2, \ldots, x_n)^T$ *as a first* approximation to the exact solution of a system. For convenience, let us denote the first approximation vector by $\left(x_1^1, x_n^1, \ldots, x_n^1\right)$ got after taking (0, 0 , …, 0) as an initial starting vector. Substituting this first approximation in the right-hand size of system, we obtain the second approximation to the given system in the form

$$\left.\begin{aligned} x_1^2 &= \frac{b_1}{a_{11}} - \frac{a_{12}}{a_{11}} x_2^1 - \ldots - \frac{a_{in}}{a_{11}} x_n^1 \\ x_2^2 &= \frac{b_2}{a_{22}} - \frac{a_{21}}{a_{22}} x_1^1 - \ldots - \frac{a_{2n}}{a_{22}} x_n^1 \\ x_n^2 &= \frac{b_n}{a_{nn}} - \frac{a_{n1}}{a_{nn}} x_1^1 - \ldots - a \frac{{}_{n(n-1)}}{a_{nn}} x_{n-1}^1 \end{aligned}\right\}$$

The second approximation is substituted into the right-hand side of *Eqs.* and obtain the third approximation and so on. This process is repeated and $(r+1)$ the approximation is calculated from

$$\left.\begin{aligned} x_1^{(r+1)} &= \frac{b_1}{a_{11}} - \frac{a_{12}}{a_{11}} x_2^{(r)} - \ldots\ldots - \frac{a_{in}}{a_{11}} x_n^{(r)} \\ x_2^{(r+1)} &= \frac{b_2}{a_{22}} - \frac{a_{21}}{a_{22}} x_1^{(r)} - \ldots\ldots - \frac{a_{2n}}{a_{22}} x_n^{(r)} \\ x_n^{(r+1)} &= \frac{b_n}{a_{nn}} - \frac{a_{n1}}{a_{nn}} x_1^{(r)} - \ldots\ldots - a_{\frac{n(n-1)}{a_{nn}} x_{n-1}^{(r)}} \end{aligned}\right\}$$

Briefly, we can rewrite $Eqs.$ as

$$x_i^{r+1} = \frac{b_i}{a_{ii}} - \sum_{\substack{j=1 \\ j \neq i}}^{n} \frac{a_{ii}}{a_{ii}} x_j^r, \; r = 1,2,\ldots, i = 1\,,2,\ldots,n$$

This method is due to Jacobi and is called Jacobi's iterative method. It is also known as method of simultaneous displacement, since no element of x_i^{r+1} is used in this iteration until every element is computed.

A sufficient condition for convergence of the iterative solution to the exact solution is

$$|a_{ii}| > \sum_{\substack{j=1 \\ j \neq i}}^{n} |a_{ij}|, \; i = 1\,,2,\ldots,n$$

When this condition (diagonal dominance) is true, Jacobi method converges.

Example-6

Find the solution to the following system of equations

$83x + 11y - 4z = 95$

$7x + 52y + 13z = 104$

$3x + 8y + 29z = 71$

Using Jacobi iterative method for the first five iterations.

Solution

A t first, we rewrite the given system in the form

$$\left.\begin{aligned} x &= \frac{95}{83} - \frac{11}{83}y + \frac{4}{83}z \\ y &= \frac{104}{52} - \frac{7}{52}x - \frac{13}{52}z \\ z &= \frac{71}{29} - \frac{3}{29}x - \frac{8}{29}y \end{aligned}\right\}$$

Taking the initial starting of solution vector as $(0,0,0)^T$, from $Eq.$ we have the first approximation as

$$\begin{pmatrix} x^{(1)} \\ y^{(1)} \\ z^{(1)} \end{pmatrix} = \begin{pmatrix} 1.1446 \\ 2.0000 \\ 2.4483 \end{pmatrix}$$

Now, using Eq. , the second approximation is compute from the equations

$$\left.\begin{aligned} x^2 &= 1.1446 - 0.1325y^1 + 0.0482_z^1 \\ y^z &= 2.0 - 0.1346_x^1 - 0.25_z^1 \\ z^2 &= 2.4483 - 0.1035x^1 - 0.2759y^1 \end{aligned}\right\}$$

Substituting Eq. into Eq we get the second approximation as

$$\begin{pmatrix} x^{(1)} \\ y^{(1)} \\ z^{(1)} \end{pmatrix} = \begin{pmatrix} 0.9976 \\ 1.2339 \\ 1.7424 \end{pmatrix}$$

GAUSS-SEIDEL ITERATION METHOD

It is another well-knows iterative method for solving a system of linear equations of the form of system. In Jacobi method, the $(r+1th$ approximation to the system is given by $s.$, from which we can observe that no element of the form of system . In Jacobi method, the $(r+1)$ th approximation to the system is given by $Eqs.$ from which we can observe that no element of x_i^{r+1} replaces x_i^r entirely for the next cycle of computation.

However, in Gauss-Seidel method, the corresponding elements of x_i^{r+1} replaces those of x_i^r as soon as they become available. Hence, it is called the method of successive displacement. For illustration, consider the system In Gauss-Seidel iteration, the $(r+1)th$ approximation or iteration is computed from

$$\left.\begin{aligned} x_1^{(r+1)} &= \frac{b_1}{a_{11}} - \frac{a_{12}}{a_{11}} x_2^{(r)} - \ldots\ldots - \frac{a_{in}}{a_{11}} x_n^{(r)} \\ x_2^{(r+1)} &= \frac{b_2}{a_{22}} - \frac{a_{21}}{a_{22}} x_1^{(r)} - \ldots\ldots - \frac{a_{2n}}{a_{22}} x_n^{(r)} \\ x_n^{(r+1)} &= \frac{b_n}{a_{nn}} - \frac{a_{n1}}{a_{nn}} x_1^{(r)} - \ldots\ldots - a_{\frac{n(n-1)}{a_{nn}} x_{n-1}^{(r)}} \end{aligned}\right\}$$

Thus, the general procedure can be written in the following compact from

$$x_i^{r+1} = \frac{b_i}{a_{ii}} - \sum_{j=1}^{i-1} \frac{a_{ii}}{a_{ii}} x_j^{r+1} - \sum_{j=i+1}^{n} \frac{a_{ii}}{a_{ii}} x_j^r$$

For all $i = 1, 2, \ldots, n$ and $r = 1, 2, \ldots$

To describe system; in the first equation, we substitute the $r-th$ approximation into the right-hand side and denote the result by x_1^{r+1}. In the second equation, we substitute $x_1^{r+1}, x_3^r, \ldots, x_n^r$ and denote the result by x_2^{r+1}*.In the thirrd equation, we* substitute $(x_i^{r+1}, x_2^{r+1}, x_4^r, \quad \ldots x_n^r)$ and denote the result. For illustration, we consider the following example

Example 7

Find the solution of the following system

$$x_1 - \frac{1}{4}x_2 - \frac{1}{4}x_3 = \frac{1}{2}$$

$$-\frac{1}{4}x_1 + x_2 - \frac{1}{4}x_4 = \frac{1}{2}$$

$$\frac{1}{4}x_1 + x_3 - \frac{1}{4}x_4 = \frac{1}{4}$$

$$-\frac{1}{4}x_2 - \frac{1}{4}x_3 + x_4 = \frac{1}{4}$$

Using Gauss-Seidel method and perform the first five iterations.

$$\left.\begin{aligned} x_1 &= 0.5 + 0.25x_2 + 0.25x_3 \\ x_2 &= 0.5 + 0.25x_1 + 0.25x_4 \\ x_3 &= 0.25 + 0.25x_1 + 0.25x_4 \\ x_4 &= 0.25 + 0.25x_2 + 0.25x_3 \end{aligned}\right\}$$

Taking $x_2 = x_3 = x_4 = 0$ on the right-hand side of the first equation of system (1), we get $x_1^1 = 0.5$. Taking $x_3 = x_4 = 0$ and the current value of x_i, we get

$$x_2^1 = 0.5 + (0.25)(0.5) + 0 = 0.625$$

From the second equation of system (1). Further, we take $x_4 = 0$ and the x_3, the fourth equation

$$x_3^1 = 0.25 + (0.25)(0.5) + 0 = 0.375$$

From the third equation of system. Now, using the current values of x_2 and x_3, the fourth equation of system (1) gives

$$x_4^1 = 0.25 + (0.25)(0.625) + (0.25)(0.375) = 0.5$$

The Gauss-Seidel iterations for the given set of equations can be written as

$$x_1^{r+1} = 0.5 + 0.25x_2^r + 0.25x_3^r$$

$$x_2^{r+1} = 0.5 + 0.25x_1^{r+1} + 0.25x_4^r$$

$$x_3^{r+1} = 0.25 + 0.25x_1^{r+1} + 0.25x_4^r$$

$x_4^{r+1} = 0.25 + 0.25x_2^{r+1} + 0.25x_3^r$

Now, by Gauss-Seidel procedure, the second and subsequent approximations can be obtained and the sequence of the first five approximations by table.

The Relaxation Method

This method is an iterative method and is due to South well. To explain the details, consider again the system of equations .Let

$$X^p = \left(x_1^p, x_2^p, \ldots., x_n^p\right)^T$$

Be the solution vector obtained iteratively after $p-th$ iteration. If R_i^p denoted the residual of the $i-th$ equation of system that is of

$$a_{i1} + a_{i2}x_2 + \ldots.. + a_{in}x_n = b_i$$

Defined by

$$R_i^p = b_i - a_{il}x_1^p - a_{i2}x_2^p - \ldots.. - a_{in}x_n^p$$

Then, we can improve the solution vector successively by reducing the largest residual to zero at that iteration. This is the basic idea of relaxation method.

To achieve the fast convergence of the procedure, we take all terms to one side and then reorder the equations so that the largest negative coefficient in the equations appear on the diagonal. Now, if at any iteration, R_i is the largest residual in magnitude, then we give an increment

$$dx_i = \frac{R_i}{a_{ii}}$$

To x_i, a_{ii} being the coefficient of x_i In other words, we change $x_i to\left(x_i + dx_i\right)$ to relax R_i, that is to reduce R_i to zero. We shall illustrate this procedure through an example.

Example 8

Solve the system of equations

$6x_1 - 3x_2 + x_3 = 11$

$2x_1 + x_2 - 8x_3 = -15$

$x_1 - 7x_2 + x_3 = 10$

By the relaxation method, starting with the vector(0, 0, 0).

Solution. At first, we transfer all the terms to the right-hand side and recorder the equations, so that the largest coefficient in the equation appear the diagonal. Thus, we get

$$\left.\begin{aligned} 0 &= 11 - 6x_1 + 3x_2 - x_3 \\ 0 &= 10 - x_1 + 7x_2 - x_3 \\ 0 &= -15 - 2x_1 - x_2 + 8x_3 \end{aligned}\right\}$$

After interchanging the second and third rows. Starting with the initial solution vector (0, 0, 0), that is taking $x_1 = x_2 = x_3 = 0,$ we find the residuals $R_1 = 11, R_2 = 10, R_3 = -15$ of which the largest residual in magnitude is $R_3.$ it indicates that the third equation has more error and needs immediate attentions for improvement. Thus, we introduce a change, $dx_3\, in\, x_3$ which is obtained from the formula

$$dx_3 = -\frac{R_3}{a_{33}} = \frac{15}{8} = 1.875$$

Similarly, we find the new residuals of large magnitude and relax it to zero, and so on. We shall continue this process, until all the residuals are zero or very small.

At this stage, we observe that all the residual $R_1\, R_2\, and\, R_3$ are small enough and therefore we may take the corresponding values of x_1 at this iteration as the solution. Hence, the numerical solution to the given system is given by

$x_1 = 1.0017$, $x_2 = -1.0$, $x_3 = 2.0$

Matrix Inversion

Consider a system of equations in the form

$$[A](X)=(B)$$

One way of writing its solution is in the form

$$(X)=[A]^{-1}(B)$$

Thus, the solution to the system can also be obtained if the inverse of the coefficient matrix [A] is known. Alternatively, if the product of two square matrices is an identity matrix, that is, if

[A] [B] = [I]

Then, [B] $=[A]^{-1}\, and [A]=[B]^{-1}$

Every square nonsingular matrix will have an inverse. Gauss elimination and Gauss-Jordan are popular among many methods available for finding the inverse of a matrix.

Gaussian Elimination Method

In this method, if *A is given* matrix, for which we have to find the inverse; at first, we place an identity matrix, whose order is same that of *A*, adjacent to A which we call an augmented matrix. Then the inverse of A is computed in two stages. In the first stage, A is converted into an upper triangular from, using Gaussian elimination method as discussed in section. In the second stage, the above upper triangular matrix reduced to an identity matrix by row transformations. All these operations are also performed on the adjacently placed identity matrix. Finally, when A is transformed into an identity matrix, the adjacent matrix gives the inverse of *A*. In order to increase the accuracy of the result, it is essential to employ partial pivoting. To understand the sequence of the step involved. We consider an example.

Example 9

Use the Gaussian elimination method to find the inverse of the matrix

$$A = \begin{bmatrix} 1\,1\,1 \\ 4\,3\,{-1} \\ 3\,5\,3 \end{bmatrix}$$

Solution. At first, we place an identity matrix of the same order adjacent to the given matrix. Thus, the augmented matrix can be written as

$$\left[\begin{array}{c|c} 1\,1\,1 & 100 \\ 4\,3\,{-1} & 010 \\ 3\,5\,3 & 001 \end{array}\right]$$

Stage 1: (reduction to upper triangular form): Let , and denote the first, second and third rows of a matrix. In the first column of Eq. (1), 4 is the largest element, thus interchanging to bring the pivot element 4 to the place of we have the augmented matrix in the for,

$$\left[\begin{array}{c|c} 4\;3\;{-1} & 01\;0 \\ 1\,1\,1 & 100 \\ 3\;5\;3 & 001 \end{array}\right]$$

Divide

$$\left[\begin{array}{c|c} 1\,\frac{3}{4}\,-\frac{1}{4} & 0\,\frac{1}{4}\,0 \\ 111 & 100 \\ 353 & 001 \end{array}\right]$$

Perform $R_2 - R_1, wich\, gives$

$$\left[\begin{array}{c|c} 1\,\frac{3}{4}\,-\frac{1}{4} & 0\,\frac{1}{4}\,0 \\ 0\,\frac{1}{4}\,\frac{5}{4} & 1\,-\frac{1}{4}\,0 \\ 353 & 0\,01 \end{array}\right]$$

Perform $R_3 - 3R_1 \rightarrow R_3 \; in\, Eq.$ which yields

$$\left[\begin{array}{ccc|ccc} 1 & \frac{3}{4} & -\frac{1}{4} & 0 & \frac{1}{4} & 0 \\ 0 & \frac{1}{4} & \frac{5}{4} & 1 & -\frac{1}{4} & 0 \\ 3 & \frac{11}{4} & \frac{15}{4} & 0 & -\frac{3}{4} & 1 \end{array}\right]$$

Now, looking at the second column for the pivot, the max $(1/4, 11/4)$ is $11/4$. Therefore, we interchanging $R_2 \; and \; R_3 \; in\, Eqs$ and get

$$\left[\begin{array}{ccc|ccc} 1 & \frac{3}{4} & -\frac{1}{4} & 0 & \frac{1}{4} & 0 \\ 0 & \frac{11}{4} & \frac{15}{4} & 0 & -\frac{3}{4} & 1 \\ 3 & \frac{1}{4} & \frac{5}{4} & 1 & -\frac{1}{4} & 0 \end{array}\right]$$

Now, divide R_2 by the pivot $a_{22} = 11/4$, and obtain

$$\left[\begin{array}{ccc|ccc} 1 & \frac{3}{4} & -\frac{1}{4} & 0 & \frac{1}{4} & 0 \\ 0 & 1 & \frac{15}{4} & 0 & -\frac{3}{4} & \frac{4}{11} \\ 3 & \frac{1}{4} & \frac{5}{4} & 1 & -\frac{1}{4} & 0 \end{array}\right]$$

Performing $R_3 - (1/4)R_2 \rightarrow R_3 \; in\, yeilds$

$$\left[\begin{array}{ccc|ccc} 1 & \frac{3}{4} & -\frac{1}{4} & 0 & \frac{1}{4} & 0 \\ 0 & 1 & \frac{15}{11} & 0 & -\frac{3}{11} & \frac{4}{11} \\ 0 & 0 & \frac{10}{11} & 1 & -\frac{2}{11} & -\frac{1}{11} \end{array}\right]$$

Finally, we divide R_3 by (10 / 11), thus getting an upper triangular form

$$\left[\begin{array}{ccc|ccc} 1 & \frac{3}{4} & -\frac{1}{4} & 0 & \frac{1}{4} & 0 \\ 0 & 1 & \frac{15}{11} & 0 & -\frac{3}{11} & \frac{4}{11} \\ 0 & 0 & 1 & \frac{1}{10} & -\frac{1}{5} & -\frac{1}{10} \end{array}\right]$$

Stage II (Reduction to an identity matrix) : Multiply $R_3\ by -$ ¼ and 15/11 respectively and subtract if from $R_1\ and\ R_2$ of . we get

$$\left[\begin{array}{ccc|ccc} 1 & \frac{3}{4} & 0 & \frac{11}{40} & \frac{1}{5} & -\frac{1}{40} \\ 0 & 1 & 0 & -\frac{3}{2} & 0 & \frac{1}{2} \\ 0 & 0 & 1 & \frac{11}{10} & -\frac{1}{5} & -\frac{1}{10} \end{array}\right]$$

Finally, performing $R_1 -$ (¾) $R_2 \to R_1\ in\ Eq.$ we obtain

$$\left[\begin{array}{ccc|ccc} 1 & 0 & 0 & \frac{7}{5} & \frac{1}{5} & -\frac{2}{5} \\ 0 & 1 & 0 & -\frac{3}{2} & 0 & \frac{1}{2} \\ 0 & 0 & 1 & \frac{11}{10} & -\frac{1}{5} & -\frac{1}{10} \end{array}\right]$$

Thus, we have

$$A^{-1}\begin{bmatrix} \frac{7}{5} & \frac{1}{5} & -\frac{2}{5} \\ -\frac{3}{2} & 0 & \frac{1}{2} \\ \frac{11}{10} & -\frac{1}{5} & -\frac{1}{10} \end{bmatrix}$$

We can easily check i.e, $[A][A^{-1}]=[I]$.

Gauss-Jordan Method

This method is similar to Gaussian elimination method, with the essential difference that the stage / of reducing the given matrix to upper triangular form is not needed. However, the given matrix can be directly reduced to an identity matrix using elementary row transformations. This technique is illustrated in the following example.

Example 10

Find the inverse of

$$A = \begin{bmatrix} 1 & 1 & 1 \\ 4 & 3 & -1 \\ 3 & 5 & 3 \end{bmatrix}$$

By Gauss-Jordan method.

Solution Let R_1, R_2 *and* R_3 denote the first, second and third rows of a matrix.We place an identity matrix adjacent to the given matrix as a first step and the resulting augmented is given by

$$\left[\begin{array}{ccc|ccc} 1 & 1 & 1 & 1 & 0 & 0 \\ 4 & 3 & -1 & 0 & 1 & 0 \\ 3 & 5 & 3 & 0 & 0 & 1 \end{array}\right]$$

Performing $R_2 - 4R_1 \to R_2$, we get

$$\left[\begin{array}{ccc|ccc} 1 & 1 & 1 & 1 & 0 & 0 \\ 4 & -1 & -5 & -4 & 1 & 0 \\ 3 & 5 & 3 & -3 & 0 & 1 \end{array}\right]$$

Carrying out further operations $R_2 + R_1 \to R_1 \; and \; R_3 + 2R_2 \to R_3$, we arrive at

$$\left[\begin{array}{ccc|ccc} 1 & 0 & -4 & -3 & 1 & 0 \\ 4 & -1 & -5 & -4 & 1 & 0 \\ 0 & 0 & -10 & 11 & 2 & 1 \end{array}\right]$$

Now, dividing the third row by – 10, we get

$$\left[\begin{array}{ccc|ccc} 1 & 0 & -4 & -3 & 1 & 0 \\ 0 & -1 & -5 & -4 & 1 & 0 \\ 0 & 0 & 1 & \frac{11}{10} & -\frac{1}{5} & -\frac{1}{10} \end{array}\right]$$

Further, we perform $R_1 + 4R_3 \to R_1, and \; R_2 + 5R_3 \to R_2$ to get

$$\left[\begin{array}{ccc|ccc} 1 & 0 & 0 & \frac{7}{5} & \frac{1}{5} & \frac{2}{5} \\ 0 & -1 & 0 & \frac{3}{2} & 1 & -\frac{1}{2} \\ 0 & 0 & 1 & \frac{11}{10} & -\frac{1}{5} & -\frac{1}{10} \end{array}\right]$$

Finally, multiplying $R_2 \; by -1$, we obtain

$$\left[\begin{array}{ccc|ccc} 1 & 0 & 0 & \frac{7}{5} & \frac{1}{5} & -\frac{2}{5} \\ 0 & 1 & 0 & -\frac{3}{2} & 0 & -\frac{1}{2} \\ 0 & 0 & 1 & \frac{11}{10} & -\frac{1}{5} & -\frac{1}{10} \end{array}\right]$$

Hence, we have

$$A^{-1} = \begin{bmatrix} \frac{7}{5} & \frac{1}{5} & -\frac{2}{5} \\ -\frac{3}{2} & 0 & \frac{1}{2} \\ \frac{11}{10} & -\frac{1}{5} & \frac{1}{10} \end{bmatrix}$$

It can be easily verified that [A] $\left[A^{-1}\right]=[I]$.

CURVE FITTING

Introduction

In the science and engineering , we often come across experimental observations involving two variable such as, concentration and time, pressure and temperature, load and deflection, etc. (in general x and y). The important question is: can we find a law or a relation connecting them. Suppose , we wish to find a relation connecting load p Kg and the deflection of a beam y cm; given the maximum deflection y for various loads p as shown in fig. Typically, the figure indicates, a linear relation of the form $y=mx+c.$ In another situation the graph may indicate a parabolic or an exponential curve. In fact, Kepler formulated the third law of planetary motion, T = , from observed data pairs (x, t) , of various planets, where T is the period , and x is the mean distance of the planet from the sun. In general, the problem of finding an equation of an approximating curve , with passes through as many point as possible , is called curve fitting. Hence. The basic problem is to find an equation of the curve of the best fit.

The method of group averages, the least square method and method of moments are some of the well=known methods based on the principles of elementary statistics. Amongst them, the least square method gives a unique best fit and is highly recommended.

METHOD OF GROUP AVERAGES

Suppose we have n sets of observation in an experiment $(x_1, y_1), (x_2, y_2), \ldots\ldots..(x_n, y_n)$. Assume that a straight line

$y = mx + c$

First this data very closely. Since contains two constant m and c, we require two equations in m and c to determine these constants.

From the given data, when we note that, observed value of But, from the assumed empirical law, the expected value Now, we define the residual, e, as the difference between the observed value of y and its expected value fro the same and write

Similarly, for other sets of data, we construct residuals as

$$e_2 = y_2 - (mx_2 + c)$$

$$e_n = y_n - (mx_n + c)$$

These expression clearly indicate that some of the residuals may be positive and some of them may be negative . However, the method of group averages state that the sum of all the residuals is zero. That's is,

$$\sum_{i=1}^{n} e_1 = [y_1 - (mx_1 + c) + [y_2 - (mx_2 + c)] + [y_j - (mx_j + c)] = 0$$

and

$$\sum_{i=1}^{n} e_1 = [y_{j+1} - (mx_{j+1} + c)] + [y_{j+2} - (mx_{j+2} + c)] + [y_n - (mx_n + c)] = 0$$

On simplification, and become

$$(y_1 + y_2 + \ldots + y_n) - jc - m(x_1 + x_2 + \ldots\ldots. + x_j) = 0$$

$$(y_{j+1} + y_{j+2} + \ldots\ldots + y_n) - (n - j)c - m(x_{j+1} + x_{j+2} + \ldots.. + x_n) = 0$$

Which can be rewritten as

$$\frac{y_1 + y_2 + \ldots. y_j}{j} = \frac{m(x_1 + x_2 + \ldots.. + x_j)}{j} + c$$

$$\frac{y_{j+1}+y_{j+2}+....y_n}{n-j}=\frac{m\left(x_{j+1}+x_{j+2}+.....+x_n\right)}{n-j}+c$$

These equations are sufficient to determine the unknown constant m and c. Obviously, there is no unique way of dividing the observations into two group and hence different choices will give different values for This, of course, is the major drawback of the method of group averages. We shall illustrate this method through the following examples.

Example 1

Use the method of group averages and find a curve of the form that fits the following data:

x	10	20	30	40	50	60	70	80
y	1.06	1.33	1.52	1.68	1.81	1.91	2.01	2.11

Solution

The required curve is of the form

$y = mx^n$

Taking logarithms on both sides, we get

Let $\log_{10} y = Y,\ \log_{10} m = c$ and $\log_{10} x = X,$ then Eq. becomes

$Y = nX + c$

Now, we take logarithm of the given pairs of data and divide them in two group, such that the first group contains the first four values and the remaining constitutes the second group, as follow:

Group 1

x	y	$y= \log_{10} y$	x= $\log_{10} x$
10	1.06	0.0253	1.0000
20	1.33	0.1239	1.3010
30	1.52	0.1818	1.4771
40	1.68	0.2253	1.6021

$\sum y = 0.5563 \quad \sum x = 5.3802$

Group 2

x	y	$y = \log_{10} y$	$x = \log_{10} x$
50	2.81	0.02577	1.6990
60	1.91	0.2810	1.7782
70	2.01	0.3032	1.8451
80	2.11	0.3243	1.9031

$\sum y = 1.1662 \quad \sum x = 7.2254$

Now, using the method of group averages, we determine the constant n and c from and as follows:

$$\frac{1}{4}\left(y_1 + y_2 + y_3 + y_4\right) = \frac{n}{4}\left(x_1 + x_2 + x_3 + x_4\right) + c$$

$$\frac{1}{4}\left(y_5 + y_6 + y_7 + y_8\right) = \frac{n}{4}\left(x_5 + x_6 + x_7 + x_8\right) + c$$

Substituting the values from the above tables, we get

$$\frac{0.5563}{4} = n\frac{53802}{4} + c$$

$$\frac{1.1662}{4} = n\frac{7.2254}{4} + c$$

On simplification, we have

4c + 5.3802n – 0.5563 = 0

4c + 7.2254n - 1.1662 = 0

Solving and by the method of cross multiplication we obtains

$c = -0.3055,\ n = 0.3305$

Therefore ,

$m = antilog\, c = 0.4949$

Hence, the required curve fit is

$y = 0.4949x^{0.3305}$

Example 2

Using the method of group averages, find a curve of the form that fits the following data:

x	20	30	35	40	45	50
y	10.0	11.0	11.8	12.4	13.5	14.4

Solution

Let the required equation of fit is

$y = a + bx^2$

Setting $x^2 = X$, the above equation becomes

$y = bX + a$

Now, the given data can be grouped and tabulated as shown:

Group 1

y	x	$x = x^2$
10.0	20	400
11.0	30	900
11.8	35	1225

$\sum y = 32.8 \quad \sum x = 2525$

Group II

y		
12.4	40	1600
13.5	45	2025
14.4	50	2500

To determine the unknown constant and b, using the method of group averages,

$$\frac{32.8}{3}=\frac{2525b}{3}+a$$

$$\frac{40.3}{3}=\frac{6125b}{3}+a$$

On simplification, we have

$$2525b+3a-32.8=0$$

$$6125b+3a-40.3=0$$

Solving $s.$, we get

$$b=2.083x10^{-3},\ a=9.1799$$

Hence, the required curve fit is

$$y=9.1799+2.083x10^{-3}x^2$$

Example 3

Using the method of group averages, fit a curve of the form

$$y=\frac{x}{a+bx}$$

For the following data :

x	8	10	15	20	30	40
y	13	14	15.4	16.3	17.2	17.8

Solution

Let the equation of the required of the required fit is

$$y=\frac{x}{a+bx}$$

On rewriting , we get

$$\frac{1}{y}=\frac{a+bx}{x}=\frac{a}{x}+b$$

Setting

$$\frac{1}{y}=y,\frac{1}{x}=X$$

the above equation becomes

$Y = aX + b$

Which is linear in X and y. Now, we shall tabulate the given data into two group:

Group 1

x	$x=1/x$	y	$y=1/y$
8	0.1230	13	0.769
10	0.1000	14	0.0714
15	0.0667	15.4	0.0649

$\sum X = 0.2917 \quad \sum y = 0.2132$

Group II

x	$x=1/x$	y	$y=1/y$
20	0.0500	16.3	0.0614
30	0.0333	17.2	0.0581
40	0.0250	17.8	0.0562

$\sum X = 0.1083 \quad \sum y = 0.1757$

Using the method of group averages, we have

$$\frac{02132}{3} = a\frac{0.2917}{3} + b$$

And

$$\frac{01757}{3} = a\frac{01083}{3} + b$$

On simplification, we get

$$0.2917a + 3b - 0.2132 = 0$$

$$0.1083a + 3b - 0.1757 = 0$$

We obtain

$a=0.2045,\ b=0.0512$

Hence , the required curve fit is

$$y = \frac{x}{0.2045 + 0.0512x}$$

THE LEAST SQUARES METHOD

In curve fitting for a given data, the method of least square us considered to give the best fit, with prior knowledge about the shape of the curve. Suppose $(x_1, y_1), (x_2, y_2) \ldots\ldots (x_n, y_n) be n$ set of observation in an experiment, Let us assume that

$$Y = f(x)$$

Is a relationship between x and y , When $x = x_1,$ the observed value of y is y_1 and the expected value of y from . Eqs is $f(x_1)$. Then the residual is defined by

$$e_1 = y_1 - f(x_1)$$

Similarly, we can define all other residuals $e_2, e_3 \ldots\ldots, e_n$ as

$$e_2 = y_2 - f(x_2)$$

$$e_n = y_n - f(x_n)$$

It is clear from the different of residuals that some of them may be positive and some may be negative. In order to give equal importance to both positive and negative residuals , it is better to consider the sum of the square of residuals, rather than sum of residuals as in the method of group averages . Thus, we consider

$$E = e_1^2 + e_2^2 + \ldots + e_n^2$$

$$= [y_1 - f(x_1)]^2 + [y_2 = f(x_2)]^2 + \ldots\ldots\ldots [y_n - f(x_n)]^2$$

Here, the quantity E is a measure of how well the curve $y = f(x)$ fits the given set of observations as a whole. Therefore, E will be zero if and only if all the points of observations lie on the curve given by Eq. The value of E decreases depending on the closeness of the observed data to the curve assumed. Hence, the best representative curve to the given set observations is one for which E, the sum of the squares of the residuals, is sum of the squares of the residuals is minimum. This concept is known as the principle of least squares. In general, we consider, straight line, parabola or an exponential curve for fitting the given data.

Fitting a Straight Line

Suppose $(x_1, y_1), (x_2, y_2), \ldots\ldots, (x_n y_n)$ be a set of n observations in an experiment and we wish to fit a straight line

$$y = ax + b$$

To these observations. That is, we have to determine the constant $a\ and\ b,$ using the principal of least squares

For any $x_i\,(i = 1, 2, \ldots\ldots, n)$ the expected value of y is $ax_i + b,$ while the observed value of $y\,is\,y_i$ Hence , the residuals are

$$e_1 = Y_I - (ax_i + b),\ for\ i = 1, 2, \ldots\ldots, n$$

Now, the sum of the squares of residuals is given as

$$E = \sum_{i=1}^{n} \left[y_i - (ax_i + b) \right]^2$$

Where E is a function of the parameters $a\,and\,b$. The necessary conditions for E to be minimum gives.

$$\frac{\partial E}{\partial a} = \frac{\partial E}{\partial b} = 0$$

In other words, the first condition yields

$$2x_1(y_1 - ax_1 - b) + 2x_2(y_2 - ax_2 - b) + \ldots\ldots + 2x_n(y_n - ax_n - b) = 0$$

That is, in compact form we write it as

$$a\sum_{i=1}^{n}x_i^2+b\sum_{i=1}^{n}x_i=\sum_{i=1}^{n}x_iy_i$$

Similarly , the second condition

$$\frac{\partial E}{\partial b}=0$$

Gives

$$(y_1-ax_1-b)+(y_2-ax_2-b)+\ldots\ldots+(y_n-ax_n-b)=0$$

That is,

$$a\sum_{i=1}^{n}x_i+nb=\sum_{i=1}^{n}y_i$$

Equations and are called normal equations, which when solved give the values of the constants $a\,and\,b.$

To compute *E,* the sum of the squares of the residuals, $Eq.$ can be recast as

$$\sum_{i=1}^{n}(y_1-ax_1-b)^2=\sum_{i=1}^{n}\left[y_i^2-2ax_iy_i-2by_i+(ax_i+b)^2\right]$$

Now, using Eq., we obtain a convenient formula to compute E as

$$E=\sum_{i}^{2}-\text{a}\;\Sigma x_iy_i-b\Sigma y_i$$

Example 6 Fit a parabola to the following data using the method of least squares.

x	1.0	1.2	1.4	1.6	1.8	2.0
y	0.98	1.40	1.86	2.55	2.28	3.20

Solution

Let the equation of the curve of parabolic fit is

$$y=ax^2+bx+c$$

The corresponding normal equations are

$$\left.\begin{aligned} a\sum x_i^2 + b\sum x_i^3 + c\sum x_i^2 &= \sum x_i^2 y_i \\ a\sum x_i^3 + b\sum x_i^2 + c\sum x_i &= \sum x_i y_i \\ a\sum x_i^2 + b\sum x_i + nc &= \sum y_i \end{aligned}\right\}$$

From the given data, we have Substituting these summations into normal equations , we get

$$39.9664a + 23.4b + 14.2c = 33.3568$$

$$23.4a + 14.2b + 9.0c = 19.848$$

$$14.2a + 9.0b + 6c = 12.27$$

Solving Eqs. We obtain

$a = 0.1875$, $b = 2.6239, c = -1.4471$

$$\log_{10} y = \log_{10} a + b \log_{10} x$$

Let

$\log_{10} y = Y$, $\log_{10} a = A$, $\log_{10} \mathrm{x} = \mathrm{X}$

Then the above equations reduce to

$$Y = A + bX$$

Which is linear in Y *and* X. To illustrate the method, we consider an example.

Example 7

By using the method of least squares, find a relation of the form $y = ax^b$, that first the data

x	2	3	4	5
y	27.8	62.1	110	161

Solution Let the equation of fit be,

$$y = ax^b$$

Taking logarithm on both sides, we get

$$\log_{10} y = \log_{10} a + \mathrm{b} \log_{10} x$$

Which is of form

$Y = A + bX = bX + A$

Where ,

$\log_{10} y = Y, \log_{10} a = \text{A}, \ \log_{10} x = X$

Now, the data in modified variable X and Y is

x	0.3010	0.4771	0.6021	0.6990
y	1.4440	1.7931	2.0414	2.2068

The corresponding normal equations are

$b\Sigma X_i^2 + A\Sigma X_i = \Sigma X_i Y_i$

$b\Sigma X_i + nA = \Sigma Y_i$

From the above data, we have

X	Y	XY	
0.3010	1.4440	0.4346	0.0906
0.4771	1.7931	0.8555	0.2276
0.6021	1.7931	1.2291	0.3625
0.6990	2.2068	1.5426	0.4886

$\Sigma X_i = 2.0792 \ \Sigma y_i = 7.4853 \ \Sigma X_i y_i = 4.0618 \Sigma X_i^2 = 1.1693$

Substituting these summations into normal equations we get

Solving these equations, we obtain $b = 1.9311, A = 0.8678$.

Therefore, $a = antilog\, 0.8678 = 7.3575.$

Hence, the required fir for the given data is

$y = 7.375 x^{1.9311}$

Fitting an Exponential Curve

Suppose we have n step of observations $(x_i, y_i), i = 1, 2, \ldots, n$ in an experiment and we wish to fit an exponential curve of the form,

$$y = ae^{bx}$$

To these observation. At first, we take logarithms on both sides to get

$$\log_{10} y = \log_{10} a + bx \log_{10} e$$

Let the above equation becomes

$$Y = Bx + A$$

Thus, it is equivalent to fitting a straight line. The procedure is illustrated through the following example.

Example 8

Using the principal of least squares, fit an equation of the form $y = ae^{bx}$ to the data.

x	1	2	3	4
y	1.65	2.70	4.50	7.35

Solution Let

$$y = ae^{bx}$$

Is the desired fit to the given data/ Taking logarithms both sides, we have

$$\log_{10} y = \log_{10} a + bx \log_{10} e$$

Suppose

$$\log_{10} y = Y, \quad \log_{10} a = A, \quad b \log_{10} e = B$$

Then the above equations assumes the form

$$Y = BX + A$$

The least square normal are

$$B\sum x_i^2 + A\sum x_i = \sum x_i y_i$$

$$B\sum x_i + 4A = \sum y_i$$

From the given data, we have

x	y	$y = \log_{10} y$	x^2	xy
1	1.65	0.2175	1	0.2175
2	2.70	0.4314	4	0.8628
3	4.50	0.6532	9	1.9596
4	7.35	0.8663	16	3.4652

$\sum x_i = 10$ $\sum y_i = 16.2$ $\sum y_i = 2.1684$ $\sum x_i^2 = 30$ $\sum x_i y_i = 6.5051$

Substituting these summations into normal equations we get

$30B + 10A = 6.5051$

$10B + 4A = 2.1684$

Solving these equations, we obtain B = 0.2168, A = 0 Now

B = b $\log_{10} e = 0.4343b = 0.2168,$

Therefore,

$B = 0.4992,\ a = 1$

Hence, the required equation of fit is

$y = e^{0.4992x}$

Method of Moments

Consider n set of observations $(x_i, y_i), i = 1, 2, \ldots, n$ in an experiment, such that x_i's are equally spaced. That is ,

$x_2 - x_1 = x_3 - x_2 = \ldots\ldots = x_n - x_{n-1} = \Delta x$

For such set of observation, we shall define the observed moments as follow

$$\left.\begin{aligned} \mu_1 &= The\ first\ moment = \sum y_i\, \Delta x = \Delta x \sum y_i \\ \mu_2 &= The\ second\ moment = \sum x_i y_i\, \Delta x = \Delta x \sum x_i y_i \\ \mu_3 &= The\ third\ moment = \sum x_i^2\, y_i\, \Delta x = \Delta x \sum x_i^2\, y_i \end{aligned}\right\}$$

And so on these moments are known as moments of the observed values of y. We shall also define the expected moments of the computed values of y, assuming $y = f(x)$ be an equations of the curves fitting the given data, as

$$\gamma_1 = The\ first\ moment\ of = \int y\,dx = \int f(x)\,dx$$
$$\gamma_2 = The\ second\ moment = \int xy\,dx = \int xf(x)\,dx$$
$$\gamma_3 = The\ third\ moment = \sum x_i^2\, y_i \Delta x = \Delta x \sum x_i^2\, y_i$$

The method of moments states that the moments of the observed y's are respectively equal to the expected moments. That is,

$$\mu_1 = \gamma_1, \mu_2 = \gamma_2\ \mu_3 = \gamma_3$$

And so on. The observed moments can be easily calculated from given data , while the moments of the expected y's can be computed as follows.

$$\left.\begin{aligned}
\gamma_1 &= \int y\,dx = \int_{x_1+\left(\frac{\Delta x}{2}\right)}^{x_n+\left(\frac{\Delta x}{2}\right)} f(x)\,dx \\
\gamma_2 &= \int xy\,dx = \int_{x_1+\left(\frac{\Delta x}{2}\right)}^{x_n+\left(\frac{\Delta x}{2}\right)} xf(x)\,dx \\
\gamma_3 &= \int x^2\, y\,dx = \int_{x_n+(\Delta x/2)}^{x_n+(\Delta x/2)} x^2 f(x)\,dx
\end{aligned}\right\}$$

That from the principal of method of moments, we have the following set of equations:

$$\left.\begin{aligned}
\Delta x \sum y_1 &= \int_{x_1+\left(\frac{\Delta x}{2}\right)}^{x_n+\left(\frac{\Delta x}{2}\right)} xf(x)\,dx \\
\Delta x \sum x_i\, y_i &= \int_{x_1+\left(\frac{\Delta x}{2}\right)}^{x_n+\left(\frac{\Delta x}{2}\right)} xf(x)\,dx \\
\Delta x \sum x_i^2 y_i &= \int_{x_1+\left(\frac{\Delta x}{2}\right)}^{x_n+\left(\frac{\Delta x}{2}\right)} x^2 f(x)\,dx
\end{aligned}\right\}$$

Given the form of $f(x)$ we can use Eqs to determine the unknown constant in $f(x)$. This method is illustrate d through the following examples

Given the form of $f(x)$ we can use Eqs to determine the unknown constant in $f(x)$. This method is illustrate d through the following examples

Example 9 Fit a straight line of the form to the following data by the method of moments

x	2	3	4	5
y	27	40	55	68

Solution Since we have to determine two constant for the required fit, it is enough to compute the first-two moments. In this example, we note that $\Delta x = 1$. Thus the observed moments are

$$\mu_1 = \Delta x \sum_{i=1}^{4} y_i = 1(27+40+55+68) = 190$$

$$\mu_2 = \Delta x \sum_{i=1}^{4} x_i y_i = 1(54+120+220+340) = 734$$

Now, we shall compute the expected moments for the given y. Therefore,

$$\gamma_1 = \int_{1.5}^{5.5} (ax+b)\,dx = \left[a\frac{x^2}{2} + bx \right]_{1.5}^{5.5} = 14a + 14b$$

The second moment of the expected value of y is

$$\gamma_2 = \int xy\,dx \int_{1.5}^{5.5} x(ax+b)\,dx = \left[a\frac{x^2}{2} + b\frac{x^2}{2} \right]_{1.5}^{5.5} = \frac{163}{3}a + 14b$$

Using the method of moments, we have

$$14a + 4b = 190, \frac{163}{3}a + 14b = 734$$

Solving these equations, we get $a = 12.9375, b = 22188$ hence, the required straight line fit is

$y = 12.9375x + 2.2188$

Example 10 Fit a parabola of the form $y = a + bx + cx^2$ to the data given below, using the method of moments

x	3	4	5	6	7
y	31.9	34.6	33.8	27.0	31.6

Solution Since we have to fit a parabola of the form $y = a + bx + cx^2$ to the given data, we have determine three constants. Therefore, it necessary to compute the first-three moments. In this problem x's are equally spaced and we also note that $\Delta x = 1$

Now, we compute the first moment as

$\mu_1 = \textit{The first moment of the observed } y\text{'s}$

$\Delta = x \sum y_i$

$= 1[31.9 + 34.6 + 33.8 + 27.0 + 31.6] = 158.9$

$\mu_2 = \textit{The second moment of the observed } y\text{'s}$

$\Delta = x \sum x_i\, y_i$

$= [95.7 + 138.4 + 169.0 + 162 + 221.2] = 786.3$

$\mu_3 = The\,Third\ moment\ of\ the\,observed\ y's$

$\Delta = x\sum x_i^2\, y_i$

$= [9x31.9 + 16x34.6 + 25x33.8 + 36x27 + 49x31.6] = 4206.1$

We also need to compute the moment of the expected values of y's given by $y = a + bx + cx^2$. Thus , we have

$\gamma_1 = The\ first\ moment\ of\ the\,expected\ values\ of\ y$

$$= \int y\,dx = \int_{2.5}^{7.5} \left(a + bx + cx^2\right) dx$$

$$= \left[ax + b\frac{x^2}{2} + c\frac{x^3}{3}\right]_{2.5}^{7.5}$$

$= 5a + 25b + 135.4167c$

Similarly

$$\gamma_2 = \int xy\,dx \int_{2.5}^{7.5} x\left(a + bx + cx^2\right) dx$$

$$\left[a\frac{x^2}{2} + b\frac{x^3}{3} + c\frac{x^4}{4}\right]_{2.5}^{7.5}$$

$= 25a + 135.42b + 781.25c$

and

$$\gamma_3 = \int x^2 y\,dy \int_{2.5}^{7.5} x^2\left(a + bx + cx^2\right) dx$$

$$= \left[a\frac{x^3}{3} + b\frac{x^4}{4} + c\frac{x^5}{5}\right]_{2.5}^{7.5}$$

$= 135.42a + 781.25b + 4726.562c$

From the method of moment , we have at once

$5a + 25b + 135.42c = 158.9$

25a + 135.42b + 781.25c = 786.3

135.42a + 781.25b + 4726.562c = 42016.1

Solving these equations, we obtain

$A=15.7217, b=7.9427, c=-0.8734$

Hence, the required parabolic fit to the given data is

$y = 15.7217 + 7.9427x - 0.8734x^2$

SUMMARY AND DISCUSSION

Many real-life problems in engineering give rise to a system of linear equations. For example , such systems occurs in certain applications of statistical analysis and in finding the numerical solution of partial differential equations and so on, it is therefore natural to seek efficient methods for solving these equations numerically .

The solution of the system of equation gives n unknown values $x_1, x_2 \ldots, x_m,$ which satisfy the system simultaneously, if $m > n,$ we may not be able to find a solution, in principle, which satisfy all the equations. If $m > n,$ the system usually will have an infinite number of solutions. However, in this chapter , we shall restrict to the case $m = n.$ In this case, $if\,|A| \neq,$ then the system will have a unique solution, while $|A| = 0,$ then there exist no solution.

Various numerical methods are available for finding the solution of the system of equations and they are classified as direct and iterative methods. In direct methods, we get the solution of the system after performing all the steps involved in the procedure. The direct methods consist of elimination methods and decomposition methods. In this chapter, under elimination methods, we consider, Gaussian elimination and Gauss-Jordan elimination methods. Crout's reduction also known as Cholesky's reduction is considered under decomposition methods. Under iterative methods, the initial approximate solution is assumed to be know and is improved towards the exact solution in an iterative methods. All these methods are easily adoptable to computers and can be used to solve even hundred or more simultaneous linear equations.

GAUSSIAN ELIMINATION METHOD

In the Gaussian elimination method, the solution to the system of Equations is obtained in two stages. In the first stage, the given system of equations is reduced to an equivalent upper triangle system is solved using back substitution procedure by which we obtain the solution in the order $x_n, x_{n-1}, x_{n-2}, \ldots\ldots x_2, x_1$.

Partial and full pivoting

The Gaussian elimination method fails if any one of the pivot elements becomes zero. In such a situation, we rewrite the equations in a different order to avoid zero pivots. Changing the order of equations is called *pivoting.*

We now introduce the concept of partial pivoting. In this technique, if the pivot happens to be zero, then the i- th column elements are searched for the numerically largest elements. Let the j-th row (j> i) contains this element, then we interchange the i-th equation with the j-th equation and proceed for elimination.

In full pivoting which is also known as complete pivoting, we interchange rows as as columns, such that the largest element in the matrix of the system becomes the pivot element. In this process, the position of the unknown variable also get changed. Full pivoting ,in fact, is more complicated that the partial pivoting. Partial pivoting is preferred for hand computation.

GAUSS-JORDAN ELIMINATION METHOD

This method is a variation of Gaussian elimination method, In this method, the elements above and below the diagonal are simultaneously made zero and thereby the given system is reduced to a equivalent diagonal from using elementary transformations. Then the solution of the resulting diagonals system an be readily obtained.

Sometimes, we normalize the pivot row with respect to the pivot element, before elimination. Partial pivoting is also used whenever the pivot element becomes zero.

CROUT'S REDUCTION METHOD

The method is based on the fact that the coefficient matrix of the system of equations can be decomposed into the product of two matrices [L] and [U], Where [L] is a lower-triangular matrix and [U] is an upper triangular matrix with 1' s on its main diagonal . The rules for getting [L] and [U] can be obtained from the fact

$$[L][U]=[A]$$

This method is also known as Cholesky's reduction method. This technique is widely used in the numerical solution of partial differential equations.

This method is very popular from computer programming point of view, since the storage space reserved for matrix [A] can be used to store the elements of [L] and [U] at the end of computation

It may be noted that this method fails if any in this case, the system in singular . In order to familiarize with the Crout's reduction method, we consider the following examples

JACOBI'S METHOD

Jacobi's method is an iterative method, where initial approximate solution to a given system of equations is assumed and is improved towards the exact solution in an iterative way. In general, when the coefficient matrix of the system of equations is a sparse matrix (many elements are zero), iterative methods have definite advantage over direct methods in respect of economy in computer memory. Such spare matrices arise in computing the numerical solution of partial differential equations.

This method is due to Jacobi and is called Jacobi's iterative method. It is also known as method of simultaneous displacement, since no element of is used in this iteration until every element is computed.

A sufficient condition for convergence of the iterative solution to the exact solution is

$$|a_{ii}| > \sum_{\substack{j=1 \\ j \neq i}}^{n} |a_{ij}|, \; i = 1, 2, \ldots, n$$

When this condition (diagonal dominance) is true, Jacobi method converges.

GAUSS-SEIDEL ITERATION METHOD

It is another well-knows iterative method for solving a system of linear equations of the form of system. In Jacobi method, the $(r+1th$ approximation to the system is given by · , from which we can observe that no element of the form of system . In Jacobi method, the $(r+1)$ th approximation to the system is given by $Eqs.$ from which we can observe that no element of x_i^{r+1} replaces x_i^r entirely for the next cycle of computation.

However, in Gauss-Seidel method, the corresponding elements of x_i^{r+1} replaces those of x_i^r as soon as they become available. Hence, it is called the method of successive displacement.

Matrix Inversion

Consider a system of equations in the form

$$[A](X)=(B)$$

One way of writing its solution is in the form

$$(X)=[A]^{-1}(B)$$

Thus, the solution to the system can also be obtained if the inverse of the coefficient matrix [A] is known. Alternatively, if the product of two square matrices is an identity matrix, that is, if

[A] [B] = [I]

Then, [B] $=[A]^{-1} \; and [A]=[B]^{-1}$

Every square nonsingular matrix will have an inverse. Gauss elimination and Gauss-Jordan are popular among many methods available for finding the inverse of a matrix.

Gaussian Elimination Method

In this method, if *A is given* matrix, for which we have to find the inverse; at first, we place an identity matrix, whose order is same that of *A*, adjacent to A which we call an augmented matrix. Then the inverse of A is computed in two stages. In the first stage, A is converted into an upper triangular from, using Gaussian elimination method as discussed in section. In the second stage, the above upper triangular matrix reduced to an identity matrix by row transformations. All these operations are also performed on the adjacently placed identity matrix. Finally, when A is transformed into an identity matrix, the adjacent matrix gives the inverse of *A*. In order to increase the accuracy of the result, it is essential to employ partial pivoting.

Gauss-Jordan Method

This method is similar to Gaussian elimination method, with the essential difference that the stage / of reducing the given matrix to upper triangular form is not needed. However, the given matrix can be directly reduced to an identity matrix using elementary row transformations.

Curve Fitting

In the science and engineering , we often come across experimental observations involving two variable such as, concentration and time, pressure and temperature, load and deflection, etc. (in general x and y). The important question is: can we find a law or a relation connecting them. Suppose , we wish to find a relation connecting load p Kg and the deflection of a beam y cm; given the maximum deflection y for various loads p as shown in fig. Typically, the figure indicates, a linear relation of the form $y=mx+c.$ In another situation the graph may indicate a parabolic or an exponential curve. In fact, Kepler formulated the third law of planetary motion, T = $cx^{3/2}$, from observed data pairs (x, t) , of various planets, where T is the period , and x is the mean distance of the planet from the sun. In general, the problem of finding an equation of an approximating curve , with passes through as many point as possible , is called curve fitting. Hence. The basic problem is to find an equation of the curve of the best fit.

The method of group averages, the least square method and method of moments are some of the well=known methods based on the principles of elementary statistics. Amongst them, the least square method gives a unique best fit and is highly recommended.

THE LEAST SQUARES METHOD

In curve fitting for a given data, the method of least square us considered to give the best fit, with prior knowledge about the shape of the curve. Suppose $(x_1, y_1),(x_2, y_2).......(x_n, y_n) be n$ set of observation in an experiment , Let us assume that

$$Y = f(x)$$

Is a relationship between x and y , When $x = x_1,$ the observed value of y is y_1 and the expected value of y from . Eqs is $f(x_1)$. Then the residual is defined by

$$e_1 = y_1 - f(x_1)$$

Similarly, we can define all other residuals $e_2 , e_3 ...., e_n$ as

$$e_2 = y_2 - f(x_2)$$

$$e_n = y_n - f(x_n)$$

It is clear from the different of residuals that some of them may be positive and some may be negative. In order to give equal importance to both positive and negative residuals , it is better to consider the sum of the square of residuals, rather than sum of residuals as in the method of group averages .

Fitting a Straight Line

Suppose $(x_1 , y_1),(x_2, y_2),,......,(x_n y_n)$ be a set of n observations in an experiment and we wish to fit a straight line

$$y = ax + b$$

To these observations. That is, we have to determine the constant $a \, and \, b,$ using the principal of least squares

For any $x_i \left(i = 1\,,2,\ldots\ldots,n\right)$ the expected value of y is $ax_i + b,$ while the observed value of $y \, is \, y_i$ Hence, the residuals are

$e_1 = Y_I - \left(ax_i + b\right), \; for \; i = 1,2,\ldots\ldots,n$

Fitting an Exponential Curve

Suppose we have n step of observations $\left(x_i, y_i\right), i = 1\,,2\,,\ldots\ldots,n$ in an experiment and we wish to fit an exponential curve of the form,

$y = ae^{bx}$

To these observation. At first, we take logarithms on both sides to get

$\log_{10} y = \log_{10} a + bx \log_{10} e$

Let the above equation becomes

$Y = Bx + A$

Thus, it is equivalent to fitting a straight line.

Method of Moments

Consider n set of observations $\left(x_i, y_i\right), i = 1,2,\ldots,n$ in an experiment, such that $x_i\,'s$ are equally spaced. That is ,

$x_2 - x_1 = x_3 - x_2 = \ldots.. = x_n - x_{n-1} = \Delta x$

For such set of observation, we shall define the observed moments as follow

$$\left.\begin{aligned} \mu_1 &= The\ first\ moment = \sum y_i \,\Delta x = \Delta x \sum y_i \\ \mu_2 &= The\ second\ moment = \sum x_i y_i \,\Delta x = \Delta x \sum x_i y_i \\ \mu_3 &= The\ third\ moment = \sum x_i^2 \, y_i \,\Delta x = \Delta x \sum x_i^2 \, y_i \end{aligned}\right\}$$

And so on these moments are known as moments of the observed values of y. We shall also define the expected moments of the computed values of y, assuming $y=f(x)$ be an equations of the curves fitting the given data, as

$$\gamma_1 = \text{The first moment of} = \int y\,dx = \int f(x)\,dx$$
$$\gamma_2 = \text{The second moment} = \int xy\,dx = \int xf(x)\,dx$$
$$\gamma_3 = \text{The third moment} = \sum x_i^2\, y_i \Delta x = \Delta x \sum x_i^2\, y_i$$

The method of moments states that the moments of the observed y's are respectively equal to the expected moments. That is,

$$\mu_1 = \gamma_1, \mu_2 = \gamma_2\ \mu_3 = \gamma_3$$

And so on. The observed moments can be easily calculated from given data , while the moments of the expected y's can be computed as follows.

$$\left.\begin{aligned}
\gamma_1 &= \int y\,dx = \int_{x_1+\left(\frac{\Delta x}{2}\right)}^{x_n+\left(\frac{\Delta x}{2}\right)} f(x)\,dx \\
\gamma_2 &= \int xy\,dx = \int_{x_1+\left(\frac{\Delta x}{2}\right)}^{x_n+\left(\frac{\Delta x}{2}\right)} xf(x)\,dx \\
\gamma_3 &= \int x^2 y\,dx = \int_{x_n+(\Delta x/2)}^{x_n+(\Delta x/2)} x^2 f(x)\,dx
\end{aligned}\right\}$$

That from the principal of method of moments, we have the following set of equations:

$$\left.\begin{aligned} \Delta x \sum y_1 &= \int_{x_1+\left(\frac{\Delta x}{2}\right)}^{x_n+\left(\frac{\Delta x}{2}\right)} xf(x)\,dx \\ \Delta x \sum x_i\, y_i &= \int_{x_1+\left(\frac{\Delta x}{2}\right)}^{x_n+\left(\frac{\Delta x}{2}\right)} xf(x)\,dx \\ \Delta x \sum x_i^2 y_i &= \int_{x_1+\left(\frac{\Delta x}{2}\right)}^{x_n+\left(\frac{\Delta x}{2}\right)} x^2 f(x)\,dx \end{aligned}\right\}$$

Given the form of $f(x)$ we can use Eqs to determine the unknown constant in $f(x)$. This method is illustrate d through the following examples

CHAPTER 8

INTERPOLATION

INTRODUCTION

Finite difference play an important role in numerical technique, where tabulated values of the function are available. For instance, consider a function $y = f(x)$. As x takes values $x_0, x_1, x_2, \ldots\ldots x_n$, let the corresponding values of y be $y_0, y_1, y_2, \ldots\ldots\ldots, y_n$. That is, for a given table of values, $(x_k, x_k)\, k = 0, 1, 2, \ldots, n;$ the process of estimating the value of y, for any intermediate value of y, for a given value of x, lying outside the table of values of x is known as extrapolation. It may be noted that if the function $f(x)$ is known, the value of y corresponding to any x can be readily computed to the desired accuracy. But, in practice, it may be difficult or sometimes impossible to know the function $y = f(x)$ in its exact form.

To look at a practical example, let us consider the computation of trajectory of a rocket flight, where we solve the Euler's dynamical equations of motion to compute its position and velocity vectors at specifies times during the flight. Under the same conditions suppose, we require the position and velocity vector, at some other intermediate times; we need not compute the trajectory again by solving the dynamical equations. Instead, we can use the best known interpolation technique to get the desired values,

In general, for interpolation of a tabulated function, the concept of finite difference is Important. The knowledge about various finite difference operations of their symbolic relations are very much needed to establish various interpolation formulae.

FINITE DIFFERENCE OPERATORS

Forward Differences

For a given table of values $(x_k, y_k),\ k = 0,1,2,\ldots,n$ with equally spaces abscissas of a function $y = f(x)$, we define the forward difference operator Δ as follows. The first forward difference is usually expressed as

$$\Delta y_i = y_{i+1} - y_{i,} \quad i = 0,1, ..,(n-1) \qquad (1)$$

To be explicit, we write

$$\Delta y_0 = y_1 - y_0$$

$$\Delta y_1 = y_2 - y_1$$

::

$$\Delta y_{n-1} = y_n - y_{n-1}$$

These difference are called first difference of the function y and are denoted by the symbol, $\Delta y_{i,} .Here\ ,v$ is called forward difference operator:

Similarly, the difference of the first difference are called second difference, define by

$$\Delta^2 y_0 = -\Delta y_1 - \Delta y_0, \qquad \Delta^2 y_1 = \Delta y_2 - \Delta y_1$$

Thus, in general

$$\Delta^2 y_1 = \Delta y_{i+1} - \Delta y_i \qquad (2)$$

Here Δ^2 is called the second difference operators. Thus, continuing, we can define $r-th$ differnce of y, as

$$\Delta^r y_i = \Delta^{r-1} y_{i+1} - \Delta^{r-1} y_i \qquad (3)$$

By defining a difference table as a convenient device for displaying various differences, the above defined difference can be written down systematically by constructing a differences table of values $(x_k, y_k), k = 0,1, \ldots\ldots, 6$ *as shown below*:

x	y	Δy	$\Delta^2 y$	$\Delta^3 y$	$\Delta^4 y$	$\Delta^5 y$	$\Delta^6 y$
x_0	y_0						
x_1	y_1	Δy_0	$\Delta^2 y_0$				
x_2	y_2	Δy_1	$\Delta^2 y_1$	$\Delta^3 y_0$			
x_3	y_3	Δy_2	$\Delta^2 y_2$	$\Delta^3 y_1$	$\Delta^4 y_0$	$\Delta^5 y_0$	
x_4	y_4	Δy_3	$\Delta^2 y_2$	$\Delta^3 y_2$	$\Delta^4 y_1$	$\Delta^5 y_1$	$\Delta^6 y_0$
x_5	y_5	Δy_4	$\Delta^2 y_3$	$\Delta^3 y_3$	$\Delta^4 y_2$		
x_6	y_6	Δy_5	$\Delta^2 y_4$				

This difference table is called forward difference table or diagonal difference table. Here, each difference is located in its appropriate column, midway between the elements of the previous column. It can be noted that the subscript remains constant along each diagonal of the table. The first term in the table, that is y_0 is called the leading term, while the difference $\Delta y_0, \Delta^2 y_0, \Delta^3 y_0, \ldots$ are leading differences.

Example 1: Construct a forward difference table for the following values of x and y.

x	0.1	0.3	0.5	0.7	0.9	1.1	1.3
y	0.003	0.067	0.148	0.248	0.370	0.518	0.697

Solution

x	y	Δy	$\Delta^2 y$	$\Delta^3 y$	$\Delta^4 y$	$\Delta^5 y$
0.1	0.003					
0.3	0.067	0.064				
0.5	0.148	0.081	0.017			
0.7	0.248	0.100	0.019	0.002		
0.9	0.370	0.122	0.022	0.003	0.001	
1.1	0.518	0.148	0.026	0.004	0.001	0.000
1.3	0.697	0.197	0.031	0.005	0.001	0.000

Example 2. Express $\Delta^2 y_0$ and $\Delta^3 y_0$ in terms of the value of the function y.

Solution. Nothing that each higher order difference is defined in term of the lower order difference , we have

$$\Delta^2 y_0 = \Delta y_1 - \Delta y_0 = (y_2 - y_1) - (y_1 - y_0) = y_2 - 2y_1 + y_0$$

and

$$\Delta^3 y_0 = \Delta^2 y_1 - \Delta^2 y_0 = (\Delta y_2 - \Delta y_1) - (\Delta y_1 - \Delta y_0)$$

$$= (y_3 - y_2) - (y_2 - y_1) - (y_2 - y_1) + (y_i - y_0)$$

$$= y_3 - 3y_2 + 3y_1 - y_0$$

Hence, we observe that the coefficients. Of the values of y, in the expansion of $\Delta^2 y_0, \Delta^3 y_0$ are binomial coefficients. Thus, in general, we arrive at the following result.

$$\Delta^n y_0 = y_n - C_1^n y_{n-1} + C_2^n y_{n-2} - C_3^n y_{n-3} + \ldots\ldots + (-1)^n y_0 \tag{4}$$

Example 3. Show that the value of y_n can be expressed in terms of the leading value y_0 and the leading difference $\Delta y_0, \Delta^2 y_0, \ldots\ldots, \Delta^n y_0$

Solution

We have from the forward difference table

$$\left.\begin{aligned} y_1 - y_0 &= \Delta y_0 \quad or \quad y_1 = y_0 + \Delta y_0 \\ y_2 - y_1 &= \Delta y_1 \quad or \quad y_2 = y_1 + \Delta y_1 \\ y_3 - y_2 &= \Delta y_2 \quad or \quad y_3 = y_2 + \Delta y_2 \end{aligned}\right\} \tag{5}$$

And so on, similarly

$$\left.\begin{aligned} \Delta y_1 - \Delta y_0 &= \Delta^2 y_0 \quad or \quad \Delta y_1 = \Delta y_0 + \Delta^2 y_0 \\ \Delta y_2 - \Delta y_1 &= \Delta^2 y_1 \quad or \quad \Delta y_2 = \Delta y_1 + \Delta^2 y_1 \end{aligned}\right\} \tag{6}$$

And so on. Similarly, we can write

$$\left.\begin{aligned} \Delta^2 y_1 - \Delta^2 y_0 &= \Delta^3 y_0 \quad or \quad \Delta^2 y_1 = \Delta^2 y_0 + \Delta^3 y_0 \\ \Delta^2 y_2 - \Delta^2 y_1 &= \Delta^3 y_1 \quad or \quad \Delta^2 y_2 = \Delta^2 y_1 + \Delta^3 y_1 \end{aligned}\right\} \tag{7}$$

And so on. Also from Eqs. 6 and 7, we can rewrite Δy_2 as

$$\Delta y_2 = \left(\Delta y_0 + \Delta^2 y_0\right) + \left(\Delta^2 y_0 + \Delta^3 y_0\right)$$

$$= \Delta y_0 + 2\Delta^2 y_0 + \Delta^3 y_0 \tag{8}$$

From Eqs. $(5)-(8), y_3$ can be rewritten

$$y_3 = y_2 + \Delta y_2$$

$$= \left(y_1 + \Delta y_1\right) + \left(\Delta y_1 + \Delta^2 y_1\right)$$

$$= \left(y_0 + \Delta y_1\right)\left(\Delta y_1 + \Delta^2 y_1\right)$$

$$= y_0 + 3\Delta y_0 + 3\Delta^2 y_0 + \Delta^3 y_0$$

$$= \left(1 + \Delta\right)^3 y_0 \tag{9}$$

Similarly, we can symbolically write

$$y_1 = \left(1 + \Delta\right) y_0, \quad y_2 = \left(1 + \Delta\right)^2 y_0, \quad y_3 = \left(1 + \Delta\right)^3 y_0$$

Continuing this procedure, we can show in general

$y_n = (1+\Delta)^n y_0$

Hence, we obtain

$$y_n = y_0 + C_1^n \Delta y_0 + C_2^n \Delta^2 y_0 + \dots\dots\dots + \Delta^n y_0 \tag{10}$$

Equivalently, we can also write the result as

$$y_n = \sum_{i=1}^{n} C_1^n \Delta^i y_0 \tag{11}$$

Backward Difference

For a given table of values $(x_k, y_k), k = 0,1,2,\dots\dots,n$ of a function $y = f(x)$ with equally spaced abscissas, the first backward differences are usually expressed in terms of the backward difference operator ∇ as

$$\nabla y_i = y_i - y_{i-1} i = n, \quad (n-1), \dots..1 \tag{12}$$

Thus,

$\nabla y_1 = y_1 - y_0$

$\nabla y_2 = y_2 - y_1$

$\vdots \quad \vdots \quad \vdots$

$\nabla y_n = y_n - y_{n-1}$

The difference of these difference are called second difference and they are denote by $\nabla^2 y_2, \quad \nabla^2 y_3, \dots\dots \nabla^2 y_n$. That is,

$\nabla^2 y_2 = \nabla y_2 - \nabla y_1$

$\nabla^2 y_3 = \nabla y_3 - \nabla y_2$

$\vdots \quad \vdots \quad \vdots$

$\nabla^2 y_n = \nabla y_n - \nabla y_{n-1}$

Thus, in general, the second backward difference are

$$\nabla^2 y_i = \nabla y_i - \nabla y_{i-1,} \quad i = n, \ (n-1), \dots.., 2 \tag{13}$$

While *the k -th* backward difference are

$$\nabla^k y_i = \nabla^{k-i} y_i - \nabla^{k-1} y_{i-1} \qquad i = n, (n-1), ,\ldots\ldots, k \qquad (14)$$

These backward difference can be systematically arranged for a table of values $(x_k, y_k), k = 0,1,\ldots,6$ as indicated in Table.

Table 1: Backward difference

x	y	∇y	$\nabla^2 y$	$\nabla^3 y$	$\nabla^4 y$	$\nabla^5 y$	$\nabla^6 y$
x_0	y_0						
x_1	y_1	∇y_1	$\nabla^2 y_2$				
x_2	y_2	∇y_2	$\nabla^2 y_3$	$\nabla^3 y_3$	$\nabla^4 y_4$	$\nabla^5 y_5$	
x_3	y_3	∇y_3	$\nabla^2 y_4$	$\nabla^3 y_4$	$\nabla^4 y_5$	$\nabla^5 y_6$	$\nabla^6 y_6$
x_4	y_4	∇y_4	$\nabla^2 y_5$	$\nabla^3 y_5$	$\nabla^4 y_6$		
x_5	y_5	∇y_5	$\nabla^2 y_6$	$\nabla^3 y_6$			
x_6	y_6	∇y_6					

From this table, it can be observed that the subscript remains constant along every backward diagonal.

Example 4.

Show that any value of y can be expressed in terms of y_n and its backward differences.

Solution. From Eq. (12) we have

$y_{n-1} = y_n - \nabla y_n$ and $y_{n-2} = y_{n-1} - \nabla y_{n-1}$

Also from the definition as given in Eq. (13), We get

$$\nabla y_{n-1} = \nabla y_n - \nabla^2 y_n$$

From these equation, we obtain

$$y_{n-2} = y_n - 2\nabla y_n + \nabla^2 y_n$$

Similarly, we can show that

$$y_{n-3} = y_n - 3\nabla y_n + 3\nabla^2 y_n - \nabla^3 y_n$$

Symbolically, these results can be rewritten as follows:

$y_{n-1} = (1-\nabla) y_n$, $\qquad y_{n-2} = (1-\nabla)^2 y_n$, $\qquad y_{n-3} = (1-\nabla)^3 y_n$

Thus, is general, we can write $y_{n-1} = (1-\nabla)^r y_n$

That is,

$$y_{n-r} = y_n - C_1^r \nabla y_n + C_2^r \nabla^2 y_n - \ldots\ldots + (-1)^r \nabla^r y_n \qquad (15)$$

Central Differences

In some applications, central differences notation is found to be more convenient to represent the successive difference of a function. Here, we use the symbol δ to represent central difference operator and the subscript of δy for any difference as the average of the subscripts of the two members of the difference. Thus, we write

$\delta y_{1/2} = y_1 - y_0$, $\qquad \delta y_{3/2} = y_2 - y_1$,

In general

$$\delta y_i = y_{i+(1/2)} - y_{i-(1/2)} \qquad (16)$$

Higher order difference are define as follows:

$$\delta^2 y_i = \delta y_{i+(1/2)} - \delta y_{i-(1/2)}. \qquad (17)$$

These central differences can be systematically arranged as indicated in table 3

Table 2: Central Difference

x	y	δy	$\delta^2 y$	$\delta^3 y$	$\delta^4 y$	$\delta^5 y$	$\delta^6 y$
x_0	y_0						
x_1	y_1	$\delta y_{1/2}$	$\delta^2 y_1$				
x_2	y_2	$\delta y_{3/2}$	$\delta^2 y_2$	$\delta^3 y_{3/2}$	$\delta^4 y_2$	$\delta^5 y_{5/2}$	
x_3	y_3	$\delta y_{5/2}$	$\delta^2 y_3$	$\delta^3 y_{5/2}$	$\delta^4 y_3$	$\delta^5 y_{7/2}$	$\delta^6 y_3$
x_4	y_4	$\delta y_{7/2}$	$\delta^2 y_4$	$\delta^3 y_{7/2}$	$\delta^4 y_4$		
x_5	y_5	$\delta y_{9/2}$	$\delta^2 y_5$	$\delta^3 y_{9/2}$			
x_6	y_6	$\delta y_{11/2}$					

Thus, we observe that all the odd difference have a fractional suffix and all the even difference with the same subscript lie horizontally.

The following alternative notation may also be adopted to introduce finite difference operators. Let $y = f(x)$ be a functional relation between x and y, which is also denoted by y_x. Suppose, we are given consecutive values of x differing by $h\, say\, x, x+h, x+2h, x+3h, etc.$ The corresponding value of y are $y_{x,}, y_{x+3h,} etc.$ As before, we can form the differences of these values. Thus

$$\Delta y_x = y_{x+h} - y_x = f(x+h) - f(x) \tag{19}$$

$$\Delta^2 y_x = \Delta y_{x+h} - \Delta y_x$$

Similarly

$$\nabla y_x = y_x - y_{x-h} = f(x) - f(x-h) \tag{20}$$

And

$$\delta y_x = y_{x+(h/2)} - y_{x-(h/2)} = f\left(x+\frac{h}{2}\right) - f\left(x-\frac{H}{2}\right) \tag{21}$$

Shift operator, E

Let $y = f(x)$ be a function of x, and let x taken the consecutive values $x,\ x+h, x+2h, etc$. We then define an operator E having the property

$$E f(x) = f(x+h) \tag{22}$$

Thus, when E operator on $f(x),$ the result is the next value of the function. Here, E is called the shift operator. If we apply the operator E twice on $f(x),$ we get

$$E^2 f(x) = E\left[E d(x)\right] = E\left[f(x+h)\right] = f(x+2h)$$

Thus, in general, if we apply the operator E n times on $f(x),$ we arrive at

$E^n f(x) = f(x+nh)$

In terms of new notation, we can write

$E^n y_x = y_{x+nh}$

Or

$$E^n f(x) = f(x+nh) \tag{23}$$

For all real values of n. Also, if $y_0, y_1, y_3, \ldots$ are the consecutive values of the function, y_x then we can also write

$$Ey_0 = y_i, \quad E^2 y_0 = y_2, \quad E^4 y_0 = y_4, \quad \ldots\ldots, \quad E^2 y_2 = y_4 \tag{24}$$

And so on. The inverse operator E^{-1} is define as

$E^{-1} f(x) = f(x-h)$

And similarly

$$E^{-n} f(x) = f(x-nh) \tag{25}$$

Average operator, ì

The average operator μ is defined as

$$\mu f(x) = \frac{1}{2}\left[f\left(x+\frac{h}{2}\right) + f\left(x-\frac{h}{2}\right)\right] = \frac{1}{2}\left[y_{x-(h/2)} + y_{x+(h/2)} \right] \tag{25}$$

Differential operator, D

It is known that D represent a differential operator having a property

$$\left.\begin{aligned} Df(x) &= \frac{d}{dx} f(x) = f'(x) \\ D^2 f(x) &= \frac{d^2}{dx^2} f(x) = f''(x) \end{aligned}\right\} \tag{26}$$

Having defined various difference operators $\Delta, \nabla, E, \mu \, and \, D.$ we can obtain the following relations easily :

From the definition of operators Δ and E, we have

$$\Delta y_x = y_{x+h} - y_x = Ey_x - y_x = (E-1)\,y_x$$

Therefore,

$$\Delta = E - 1 \qquad (27)$$

Following the definition of operators ∇ and E^{-1}, we have

$$\nabla y_x = y_x - y_{x-h} = y_x - E^{-1}y_x = (1 - E^{-1})\,y_x$$

Therefore,

$$\nabla = 1 - E^{-1} = \frac{E-1}{E} \qquad (28)$$

The definition of operators δ *and* E gives

$$\delta y_x = y_{x+(h/2)} - y_{x-(h/2)} = E^{1/2}y_x - E^{-1/2}y_x = (E^{1/2} - E^{-1/2})\;y_x$$

Hence,

$$\delta = E^{1/2} - E^{-1/2}$$

The definition of μ *and* E similarly yields

$$\mu y_x = \frac{1}{2}\left[y_{x+(h/2)} - y_{x-(h/2)}\right] = \frac{1}{2}(E^{1/2} - E^{-1/2})\,y_x$$

Therefore,

$$\mu = \frac{1}{2}(E^{1/2} - E^{-1/2}) \qquad (30)$$

It is know that

$$Ey_x = y_{x+h} = f(x+h)$$

Using Taylors series expansion, we have

$$Ey_x = f(x) + hf'(x) + \frac{h^2}{2!}f''(x) + \ldots$$

$$= f(x) + hDf(x) + \frac{h^2}{2!}f''(x) + \ldots.$$

$$=\left(1+\frac{hD}{1!}+\frac{h^2D^2}{2!}+.....\right)f(x)=E^{hd}y_x$$

Thus,

$$hD = logE \tag{31}$$

Example 5. Prove that

$$hd=\log(1+\Delta)=-\log(1-\nabla)=\sinh^{-1}(\mu\delta)$$

Solution. Using the standard relations (27) –(31), we have

$$hd=\log E=\log(1+\Delta)=-\log E^{-1}=-\log(1-\nabla) \tag{1}$$

Also,

$$\mu\delta=\frac{1}{2}\left(E^{1/2}+E^{-1/2}\right) \qquad \left(E^{1/2}-E^{-1/2}\right)=$$

$$\frac{1}{2}\left(E-E^{-1}\right)=\frac{1}{2}\left(e^{hd}-e^{-hd}\right)=\sinh(hD)$$

Therefore,

$$hD=\sinh^{-1}(\mu\delta)$$

Equation (1) and (2) constitute the required result.

Example 6. If Δ,∇,δ denote forward backward and central difference operator, E and μ are respectively the shift and average operators, in the analysis of data with equal spacing h, show that

$$(i)\ 1+\delta^2\mu^2=\left(1+\frac{\delta^2}{2}\right),$$

$$(ii)\,E^{1/2}=\mu+\frac{\delta}{2}$$

$$(iii)\ \Delta=\frac{\delta^2}{2}+\delta\sqrt{1+(\delta^2/4}$$

$$(iv)\,\mu\delta=\frac{\Delta E^{-1}}{2}+\frac{\Delta}{2}$$

$(v)\ \mu\delta = \frac{\Delta + \nabla}{2}$

Solutions (i) from the definition of operators, we have

$$\mu\delta = \frac{1}{2}\left(E^{1/2} + E^{-1/2}\right)\left(E^{1/2} - E^{-1/2}\right) = \frac{1}{2}\left(E - E^{-1}\right)$$

Therefore,

$$1 + \mu^2\delta^2 = 1 + \frac{1}{4}\left(E^2 - 2 + E^{-2}\right) = \frac{1}{4}\left(E + E^{-1}\right)^2$$

Also,

$$1 + \frac{\delta^2}{2} = \frac{1}{2}\left(E^{1/2} + E^{-1/2}\right) = \frac{1}{2}\left(E + E^{-1}\right)$$

From ^{Eqs} (1) and (2)t, the first result follows.

(ii) Now

$$\mu + \frac{\delta}{2} = \frac{1}{2}\ \left(E^{1/2} + E^{-1/2} + E^{1/2} - E^{-1/2}\right) = \ E^{1/2}$$

Thus, the second result is proved

(iii) we can write

$$\frac{\delta^2}{2} + \delta\sqrt{1 + \left(\delta^2 / 4\right)} = \frac{\left(E^{1/2} - E^{-1/2}\right)^2}{2}$$

$$+ \frac{\left(E^{1/2} - E^{-1/2}\right)^2 \sqrt{1 + \frac{1}{4}\left(E^{1/2} - E^{-1/2}\right)^2}}{1}$$

$$= \frac{E - 2 + E^{-1}}{2} + \frac{1}{2}\left(E^{\frac{1}{2}} - E^{-\frac{1}{2}}\right)\left(E^{1/2} + E^{-1/2}\right)$$

$$\frac{E - 2 + E^{-1}}{2} + \frac{E - E^{-1}}{2}$$

$$E = -1$$

Using Eq. (27), we get

$$E-1=\Delta$$

(iv) We have

$$\mu\delta=\frac{1}{2}\left(E^{1/2}+E^{-1/2}\right)\left(E^{1/2}-E^{-1/2}\right)=\frac{1}{2}\left(E-E^{-1}\right)$$

Now, using Eq. (27), we get

$$=\frac{1}{2}\left(1+\Delta-E^{-1}\right)=\frac{\Delta}{2}+\frac{1}{2}\left(1-E^{-1}\right)$$

$$=\frac{\Delta}{2}+\frac{1}{2}\left(\frac{E-1}{E}\right)=\frac{\Delta}{2}+\frac{\Delta}{2E}$$

(v) We can write

$$\mu\delta=\frac{1}{2}\left(E^{1/2}+E^{-1/2}\right)\left(E^{1/2}-E^{-1/2}\right)=\frac{1}{2}\left(E-E^{-1}\right)$$

Now using Eq. (27) and (28), we have

$$\mu\delta=\frac{1}{2}\left(1+\Delta-1+\nabla\right)=\frac{1}{2}\left(\Delta+\nabla\right)$$

Example 7. Show that the operation $\mu \, and \, E$ commute.

Solution From the definition of operators $\mu \, and \, E,$ we have

$$\mu\delta y_0=\mu y_1=\frac{1}{2}\left(y_{3/2}+y_{1/2}\right) \qquad (1)$$

While

$$E\mu\delta y_0=\frac{1}{2}E\left(y_{1/2}+y_{-1/2}\right)=\frac{1}{2}\left(y_{3/2}+y_{1/2}\right) \qquad (2)$$

Equating (1) and (2), we have

Therefore, the operators $\mu \, and \, E$ commute.

Newton's Forward Difference Interpolation Formula

Let $y = f(x)$ be a functional which takes values $f(x_0), f(x_0+h), f(x_0+2h), \ldots..,$ corresponding to various equispaced values of x with spacing h, $x_0, x_0+h, x_0+2h, \ldots..$ Suppose, we wish to evaluate the function $f(x)$ for a value $x_0 + ph,$ where P is any real number, then for any real number P, we have the operator E such that $E^p f(x) = f(x+ph).$ therefore , using Eq. we have

$$f(x_0+ph) = E^p f(x_0) = (1+\Delta)^p (x_0)$$

$$= \left[1 + p\Delta + \frac{p(p-1)}{2!}\Delta^2 + \frac{p(p-1)(p-2)}{3!}\Delta^3 + \ldots.\right] f(x_0)$$

That is,

$$f(x_0+ph) = f(x_0) + p\Delta f(x_0) + \frac{p(p-1)}{2!}\Delta^2 f(x_0) + \frac{p(p-1)(p-2)}{3!}\Delta^3 f(x_0) + \ldots + \frac{P(p-1)\ldots(p-n+1)}{n!}\Delta^n f(x_0) + Error$$

This is known as Newton forward difference formula for interpolation, which gives the value of $f(x_0+ph)$ in terms of $f(x_0)$ and its leading differences. This formula is also known as Newton-Gregory forward difference interpolation formula: Hence: Here $P = (x - x_0) lh.$ Equation can also be written in another alternate form as

$$y_x = y_0 + p\Delta y_0 + \frac{p(p-1)}{2!}\Delta^2 y_0 + \frac{p(p-1)(p-2)}{3!}\Delta^3 y_0 + \ldots\ldots + \frac{P(p-1)\ldots(p-n+1)}{n!}\Delta^n y_0 + Error$$

If we retain $(r+1)$ terms in, we obtain a polynomial of degree r agreeing with y_x at $x_0, x_1, \ldots.., x_r.$

This formula is mainly used for interloping the values of y near the beginning of a set of tabular values and for extrapolating values of $y,$ a short distance backward form y_0 We shall illustrate these formula by considering the following examples.

Example 8. Evaluate $f(15),$ given the following table of values:

x	10	20	30	40	50
$y = f(x)$	46	66	81	93	101

Solution We many note that $x=15$ is very near to the beginning of the table. Hence, we use Newton's forward difference interpolation formula. The forward difference are calculated and tabulated as given below:

x	$y = f(x)$	Δy	$\Delta^2 Y$	$\Delta^3 y$	$\Delta^4 y$
10	46	20			
20	66	15	-5		
30	81	12	-3	2	
40	93	8	-4	-1	-3
50	101				

We have Newton's forward difference interpolation formula as

$$y = y_0 + p\Delta y_0 + \frac{p(p-1)}{2!}\Delta^2 y_0 + \frac{p(p-1)(p-2)}{3!}\Delta^3 y_0 + \frac{P(p-1)(P-2)(p-3)}{4!}\Delta^4 y_0$$

In this example, from the above table, we have

$$x_0 = 10, y_0 = 46, \Delta y_0 = 20. \Delta^2 y_0 = -5, \Delta^3 y_0 = 2,\ \ \Delta^4 y_0 = -3$$

Let y_{15} be the value of y when x = 15, then

$$P = \frac{x - x_0}{h} = \frac{15-10}{10} = 0.5$$

Substituting these values in Eq. (1), we get

$$f(15) = y_{15} = 46 + (0.5)(20) + \frac{(0.5)(0.5-1)}{2}(-5) + \frac{(0.5)(0.5-1)(0.5-2)}{6}(2)\frac{(0.5)(0.5-1)(0.5)(0.5-2)}{24}(-3)$$

$$= 46 + 10 + 0.625 + 0.125 + 0.1172$$

Example 9 Find Newton forward difference interpolating polynomial for the following data:

x	0.1	0.2	0.3	0.4	0.5
$y = f(x)$	1.40	1.56	1.76	2.00	2.28

Solution. We shall first constructed the forward difference table to the given data as indicated below:

x	$y = f(x)$	Δy	$\Delta^2 Y$	$\Delta^3 y$	$\Delta^4 y$
0.1	1.40				
0.2	1.56	0.16			
0.3	1.76	0.20	0.04		
0.4	2.00	0.24	0.04	0.00	0.00
0.5	2.28	0.28	0.04	0.00	

Since, third and fourth leading difference are zero, we have Newton's forward difference interpolating formula as

$$y = y_0 + p\Delta y_0 + \frac{p(p-1)}{2}\Delta^2 y_0 \qquad (1)$$

In this problem, $x_0 = 0.1$, $y_0 = 1.40, \Delta y_0 = 0.16, \Delta^2 y_0 = 0.04,$ and

$$P = \frac{x-0.1}{0.1} = 10x-1$$

Substituting these values in Eq. (1) , we obtain

$$y = f(x) = 140 + (10x-1)(0.16) + \frac{(10x-1)(10x-2)}{2}(0.04)$$

That is $y = 2x^2 + x + 1.28$. This is required Newton's interpolating polynomial.

Example. 10 Estimate the missing figure in the following table:

x	1	2	3	4	5
$y = f(x)$	2	5	7	–	32

Solution. Since we are given four entries in the table, the function $y = f(x)$ can be represented by a polynomial of degree three. Using Theorem, we have

$$\Delta^3 f(x) = Constant \quad and\ \Delta^4 f(x) = 0$$

For all x. In particular, $\Delta^4 f(x_0)=0.$ Equivalently, $(E-1)^4 f(x_0)=0.$ Expanding, we have

$(E^4-4E^3+6E^2-4E+1)f(x_0)=0$

That is, $f(x_4)-4f(x_3)+6f(x_2)-4f(x_1)n+f(x_0)=0$

Using the values given in the table, we obtain $32-4f(x_3)+6x7-4x5+2=0$

Which gives $f(x_3)$, the missing value equal to 14.

Example 11. Find a cubic polynomial in x which takes on the values $-3.3.11.27.57$ and 107, when $x= ,1,2,3,4$ and 5 respectively.

Solution. Here, the observation are given at equal intervals of unit width. To determine the required polynomial, we first construct the difference table as follows:

x	$f(x)$	$\Delta f(x)$	$\Delta^2 f(x)$	$\Delta^3 f(x)$
0	-3			
1	3	6		
2	11	8	2	
3	27	16	8	6
4	57	30	14	6
5	107	50	20	6

Since the fourth and higher order difference are zero, we have required Newton's interpolation formula in the form

$$f(x_0+ph)=f(x_0)+p\Delta f(x_0)+\frac{p(p-1)}{2}\Delta^2 f(x_0)+\frac{p(p-1)(p-2)}{6}\Delta^3 f(x_0) \quad (1)$$

Here,

$$P=\frac{x-x_0}{h}=\frac{x-0}{1}=x, \ \Delta f(x_0)=6, \quad \Delta^2(x_0)=2, \ \Delta^3 f(x_0)=6$$

Substituting these values into Eq. (1), we have

$$f(x) = -3 + 6x + \frac{x(x-1)}{2}(2) + \frac{x(x-1)(x-2)}{6}(6)$$

That is, $f(x) = x^3 - 2x^2 + 7x - 3,$ is the required cubic polynomial.

Example 12 For the following table of values, estimate f

x	1	2	3	4	5	6	7	8
$y = f(x)$	1	8	27	64	125	216	343	512

Solution The value to be interpolated is at end of the table. Hence, it is appropriate to use Newton's backward interpolation formula. We shall first construct the difference table for the given data:

x	$y = f(x)$	Δy	$\Delta^2 y$	$\Delta^3 y$	$\Delta^4 y$
1	1				
2	8	7			
3	27	19	12		
4	64	37	18	6	0
5	125	61	24	6	0
6	216	91	30	6	0
7	343	127	36	6	0
8	512				

Since the fourth and higher order difference are zero, the required Newton's backward interpolation formula is

$$y_x = y_n + p\nabla y_n + \frac{p(p+1)}{2!}\nabla^2 y_n + \frac{p(p+1)(p+2)}{3!}\nabla^3 y_n$$

In this problem,

$$p = \frac{x - x_n}{h} = \frac{7.5 - 8.0}{1} = -0.5$$

And $\nabla y_n = 169, \nabla^2 y_n = 42, \nabla^3 y_n = 6$

Therefore,

$$y_{7.5} = 512 + (-0.5)(169) + \frac{(-0.5)(0.5)}{2}(42) + \frac{(-0.5)(0.5)(1.5)}{6}(6)$$

$$= 512 - 84.5 - 5.25 - 0.375$$

$$= 421.875$$

SUMMARY AND DISCUSSION

Finite difference play an important role in numerical technique, where tabulated values of the function are available. For instance, consider a function $y = f(x)$. As x takes values $x_0, x_1, x_2, \ldots..x_n$, let the corresponding values of y be $y_0, y_1, y_2, \ldots\ldots., y_n$. That is, for a given table of values, $(x_k, x_k)k = 0,1,2,\ldots,n;$ the process of estimating the value of y, for any intermediate value of y, for a given value of x, lying outside the table of values of x is known as extrapolation. It may be noted that if the function $f(x)$ is known, the value of y corresponding to any x can be readily computed to the desired accuracy. But, in practice, it may be difficult or sometimes impossible to know the function $y = f(x)$ in its exact form.

In general, for interpolation of a tabulated function, the concept of finite difference is Important. The knowledge about various finite difference operations of their symbolic relations are very much needed to establish various interpolation formulae.

Forward Differences

For a given table of values (x_k, y_k), $k = 0,1,2,\ldots,n$ with equally spaces abscissas of a function $y = f(x)$, we define the forward difference operator Δ as follows. The first forward difference is usually expressed as

$$\Delta y_i = y_{i+1} - y_{i,} \quad i = 0,1, \ldots,(n-1)$$

To be explicit, we write

$$\Delta y_0 = y_1 - y_0$$

$$\Delta y_1 = y_2 - y_1$$

$$\vdots$$

$$\Delta y_{n-1} = y_n - y_{n-1}$$

These difference are called first difference of the function y and are denoted by the symbol, $\Delta y_{i,}$ $.Here\ ,v$ is called forward difference operator:

Backward Difference

For a given table of values $(x_k, y_k), k = 0,1,2,\ldots\ldots,n$ of a function $y = f(x)$ with equally spaced abscissas, the first backward differcnces are usually expressed in terms of the backward difference operator ∇ as $\nabla y_i = y_i - y_{i-1} i = n, \quad (n-1), \ldots..1$

Thus,

$$\nabla y_1 = y_1 - y_0$$

$$\nabla y_2 = y_2 - y_1$$

$$\vdots \quad \vdots \quad \vdots$$

$$\nabla y_n = y_n - y_{n-1}$$

The difference of these difference are called second difference and they are denote by $\nabla^2 y_2, \ \nabla^2 y_3, \ldots\ldots \nabla^2 y_n$. That is,

$$\nabla^2 y_2 = \nabla y_2 - \nabla y_1$$

$$\nabla^2 y_3 = \nabla y_3 - \nabla y_2$$

$$\vdots \quad \vdots \quad \vdots$$

$$\nabla^2 y_n = \nabla y_n - \nabla y_{n-1}$$

Thus, in general, the second backward difference are

$$\nabla^2 y_i = \nabla y_i - \nabla y_{i-1}, \quad i = n, (n-1), \ldots.., 2$$

While $the k\_th$ backward difference are

$$\nabla^k y_i = \nabla^{k-i} y_i - \nabla^{k-1} y_{i-1} \quad i = n, (n-1), ,\ldots\ldots.., k$$

Central Differences

In some applications, central differences notation is found to be more convenient to represent the successive difference of a function. Here, we use the symbol δ to represent central difference operator and the subscript of δy for any difference as the average of the subscripts of the two members of the difference. Thus, we write

$$\delta y_{1/2} = y_1 - y_0, \qquad \delta y_{3/2} = y_2 - y_1,$$

In general

$$\delta y_i = y_{i+(1/2)} - y_{i-(1/2)}$$

Higher order difference are define as follows:

$$\delta^2 y_i = \delta y_{i+(1/2)} - \delta y_{i-(1/2)}$$

Newton's Forward Difference Interpolation Formula

Let $y = f(x)$ be a functional which takes values $f(x_0), f(x_0 + h), f(x_0 + 2h), \ldots..,$ corresponding to various equispaced values of x with spacing h, $x_0, x_0 + h, x_0 + 2h, \ldots..$ Suppose, we wish to evaluate the function $f(x)$ for a value $x_0 + ph,$ where P is any real number, then for any real number P, we have the operator E such that $E^p f(x) = f(x+ph)$. therefore , using Eq. we have

$$f(x_0 + ph) = E^p f(x_0) = (1+\Delta)^p (x_0)$$

$$= \left[1 + p\Delta + \frac{p(p-1)}{2!}\Delta^2 + \frac{p(p-1)(p-2)}{3!}\Delta^3 + \ldots\right] f(x_0)$$

That is,

$$f(x_0 + ph) = f(x_0) + p\Delta f(x_0) + \frac{p(p-1)}{2!}\Delta^2 f(x_0) + \frac{p(p-1)(p-2)}{3!}\Delta^3 f(x_0) + \ldots + \frac{P(p-1)\ldots(p-n+1)}{n!}\Delta^n f(x_0) + Error$$

This is known as Newton forward difference formula for interpolation, which gives the value of $f(x_0 + ph)$ in terms of $f(x_0)$ and its leading differences. This formula is also known as Newton-Gregory forward difference interpolation formula.

This formula is mainly used for interloping the values of y near the beginning of a set of tabular values and for extrapolating values of y, a short distance backward form y_0.

CHAPTER 9

DIFFERENTIAL EQUATIONS CONCEPTS FOR MATRIX ALGEBRA

INTRODUCTION

In many application of science and engineering, these has been a growing desire for numerical answer, for the design of any system and to study its performance. Therefore, in recent years greater emphasis has shifted from analytic technique to computer oriented numerical methods. The principal attraction of numerical methods is that solutions could be obtained for many problems which are not amenable to analytical treatment. The availability of the modem digital computers paved the way for the development of efficient and more general finite difference methods for solving particularly partial differential equations, [The broad classification details and analytical solutions of partial differential equations are covered in Sankara Rao (1995)].

In this chapter, we shall present some of the important and powerful finite difference method for solving parabolic partial differential equations that model heat and mass transfer in fluids and solids, unsteady behavior of fluid flow past bodies, etc For example, if the fluid is incompressible and has constant thermal conductivity *k,* the law of conservation of energy can be expressed mathematically in the form $pC_p \frac{DT}{dt} = k\nabla^2 T$ where p is the density, C_p is the specific heat of the fluid at constant pressure . T is the temperature and D / Dt is the material derivative. In Cartesian coordinates, can be written as

$$\frac{\partial T}{\partial t}+\mu\frac{\partial T}{\partial x}+v\frac{\partial T}{\partial y}+w\frac{\partial T}{\partial z}=\alpha\left(\frac{\partial^2 T}{\partial x^2}+\frac{\partial^2 T}{\partial y^2}+\frac{\partial^2 T}{\partial z^2}\right)$$

Here $\alpha=\frac{k}{pC_p};\mu, v\,and\,w\,are$ the components of the fluid at a point (x,y,z) and at time *t*. and at time *t* infect, it is a non –linear parabolic partial differential equation for fluids which are at rest, becomes

$$\frac{\partial T}{\partial t}=\left(\frac{\partial^2 T}{\partial x^2}+\frac{\partial^2 T}{\partial y^2}+\frac{\partial^2 T}{\partial z^2}\right)$$

Which is also called a diffusion equation.

The finite difference method for solving equations of the type though fairly simple is also a powerful technique for solving a variety of heat transfer and fluid flow problems in regular geometries. Most of the concepts associated with the numerical solutions of partial differential equations by finite difference methods can be best be illustrated and understood by considering a simple one. Dimensional diffusion equation in the form.

$$\frac{\partial T}{\partial t}=\alpha\frac{\partial^2 T}{\partial x^2}$$

which hold, true in some prescribed region R of the (x, t) space.

BASIC CONCEPTS IN FINITE DIFFERENCE METHODS

For the purpose of illustration, let us consider a two dimensional region as it is covered by a rectangular grid formed by two sets of lines drawn parallel to the coordinate axes with grid spacing $\Delta x\ and\ \Delta y\ in\ x\ and\ y$ directions respectively. The numerical values of the dependent variable are obtained at the points of intersection of the parallel lines, called mesh points or differential equations over the region of interest to derive approximately equivalent algebraic equations. The discretization consists of replacing each derivate of the partial differential equations at a mesh point by a finite difference approximation in terms of the values of the dependent variable at the mesh point and at the immediate neighboring mesh points, and boundary points. In doing so, a set of algebraic equations arise. Let the temperature T at a representation point be a function of two special coordinates x, y and time t. We adopt the following notation. Let the subscripts $i \, and \, j$ represent x and y coordinates and superscripts, n represents time. Let the mesh spacing in x and y directions are denoted by $\Delta x \, and \, \Delta y$ and the time step by Δt. Thus T(x, y, t) can be represented by T($i\Delta x, j\Delta y, n\Delta t) = T_i^n j$. With this notations, Let us assume that the function T and its derivates are continuous. Then the finite difference approximations to derivates can be obtained from Taylor's series expansions. For example, the Taylor's series expansion of $T_{i+1} j$ about the grid point (i, j) gives

$$T_{i+1,j} = T_{i,j} \left[\Delta x \frac{\partial T}{\partial x} + \frac{(\Delta x)^2}{2} \frac{\partial^2 T}{\partial x^2} + \frac{(\Delta x)^3}{6} \frac{\partial^3 T}{\partial x^3} + \frac{(\Delta x)^4}{24} \frac{\partial^4 T}{\partial x^4} + higher\, order\, terms \right]_{i,j}$$

Similarly,

$$T_{i+1,j} = T_{i,j} \left[\Delta x \frac{\partial T}{\partial x} - \frac{(\Delta x)^2}{2} \frac{\partial^2 T}{\partial x^2} + \frac{(\Delta x)^3}{6} \frac{\partial^3 T}{\partial x^3} - \frac{(\Delta x)^4}{24} \frac{\partial^4 T}{\partial x^4} + higher\, order\, terms \right]_{i,j}$$

Solving for $\partial T / \partial x$, gives

$$\left(\frac{\partial T}{\partial x} \right)_{i,j} = \frac{T_{i+1} - T_{i,j}}{\Delta x} + \mathrm{O}(\Delta \mathrm{x})$$

While Eq. gives

$$\left(\frac{\partial T}{\partial x} \right)_{i,j} = \frac{T_{ij} - T_{i-1,j}}{\Delta x} + \mathrm{O}(\Delta \mathrm{x})$$

Equations is the forward difference approximation to the derivative $\partial T \;/\; \partial x$, while Eq. is its backward difference approximation. Both approximations have a truncation error of order Δx, thst is they are first order accurate. Subtracting Eq. from Eq. we obtain

$$\left(\frac{\partial T}{\partial x} \right)_{i,j} = \frac{T_{i+1j} - T_{i-1,j}}{2\Delta x} + \mathrm{O}\left[(\Delta \mathrm{x})^2 \right]$$

This is a central difference approximation to the derivative $\partial T / \partial x$, and is second order accurate, in the sense that it has a truncation error of order $0(\Delta x^2)$.

The central difference approximation to a second order partial derivative $\partial^2 T / \partial x^2$ can be similarly obtained by adding $Eqs.$ and. Thus

$$\left(\frac{\partial^2 T}{\partial x^2}\right)_{i,j} = \frac{T_{i+1j} - 2T_{i,j} + T_{i-1,j}}{\Delta x^2} + O(\Delta x)^2$$

Which is second order accurate. Similar expressions can be written for y derivatives.

$$\left(\frac{\partial^2 T}{\partial y^2}\right)_{i,j} = \frac{T_{i+1j} - 2T_{i,j} + T_{i-1,j}}{\Delta y^2} + O(\Delta y^2)$$

The expression for mixed derivatives can be obtained by differentiating with respect to each variable to each variable in turns. Thus, for example,

$$\left(\frac{\partial^2 T}{\partial x \partial y}\right)_{i,j} = \frac{\partial}{\partial x}\left(\frac{\partial T}{\partial y}\right)_{i,j} = \frac{\left(\frac{\partial T}{\partial y}\right)_{i+1,j} - \left(\frac{\partial T}{\partial y}\right)_{i-1,j}}{2\Delta x}$$

$$= \frac{\left(T_{i+1,j+1} - T_{i+1,j-1}\right)/(2\Delta y) - \left(T_{i-1,j+1} - T_{i-1,j-1}\right)/(2\Delta y)}{2\Delta y}$$

Therefore,

$$\left(\frac{\partial^2 T}{\partial x \partial y}\right)_{i,i} = \frac{T_{i+1,j+1} - T_{i+1,j-1} - T_{i-1,j+1} - T_{i-1,j-1}}{4\Delta x \Delta y}$$

Proceeding in a similar manner, the central difference approximation to the third derivative is found to be

$$\left(\frac{\partial^3 T}{\partial x^3}\right)_{i,j} = \frac{T_{i+2,j} - 2T_{i+1,j} + 2T_{i-1,j} - T_{i-2,j}}{2(\Delta x)^3}$$

Similar approximations can be obtained even to higher order derivatives. It may be observed that, as the other of the derivative increases, the number of adjacent nodes involved in the approximation also increases. When we are computing adjacent to the boundaries, the usefulness of these approximations become restricted. In fact, while computing near the boundarie, one sided formulae involving more adjacent points become necessary. For instance, to implement derivative boundary conditions, a second order accurate difference formula can be derived involving simultaneously the nodes $(i-2,j),(i-1,j)\,and\,(i,j)$ in the form

$$\left(\frac{\partial T}{\partial x^2}\right)_{i,j}=\frac{\alpha T_{ij}+bT_{i-1,j}+cT_{i-2j}}{\Delta x}$$

Where the coefficient a,b,c can be obtained from Taylor's series expansion of $T_{i-2,j}\,and\,T_{i-1j}$ as shown below:

$$T_{i-2,j}=T_{ij}-(2\Delta x)\left(\frac{\partial T}{\partial x}\right)_{i,j}+\frac{(2\Delta x)^2}{2}\left(\frac{\partial^2 T}{\partial x^2}\right)_{i,j}-\frac{(2\Delta x)^3}{6}\left(\frac{\partial^3 T}{\partial x^3}\right)+H.O.T$$

$$T_{i-1,j}=T_{ij}-\Delta x\left(\frac{\partial T}{\partial x}\right)_{i,j}+\frac{(\Delta x)^2}{2}\left(\frac{\partial^2 T}{\partial x^2}\right)_{i,j}-\frac{(\Delta x)^3}{6}\left(\frac{\partial^3 T}{\partial x^3}\right)+H.O.T$$

Thus, we have

$$aT_{ij}+bT_{i-1j}+cT_{i-2,j}=(a+b+c)T_{ij}-\Delta x(2c+b)\left(\frac{\partial T}{\partial x}\right)_{i,j}+\frac{(\Delta x)^2}{2}(4c+b)\left(\frac{\partial^2 T}{\partial x^2}\right)_{i,j}+O\left(\Delta x^3\right)$$

Comparing Eqs , we get

$$a+b+c=0,\;\; b+2c=-1,\;\; b+4c=0$$

Solving these equations, we obtain

$$a=\frac{3}{2},b=-\frac{4}{2},c=\frac{1}{2}$$

Thus,

$$\left(\frac{\partial T}{\partial x}\right)_{i,j}=\frac{3T_{ij}-4T_{i-1,j}+T_{i-2j}}{2\Delta x}O\left(\Delta x^2\right)$$

Similarly, one –sided formulae can be derived even from higher order derivatives

Alternatively, in, we have defined various finite difference operator $\Delta,\nabla\,\delta,E,\mu\,and\,D$ and obtained various relations connecting them. Those relations can also be used to derive finite difference approximations to derivatives of any order to the desired accuracy

Thus, the application of a finite difference method to a give physical problems consist of the following three basic step:

Division of spatial domain into an orthogonal computational grid.

Discretization of the governing equations and boundary conditions in space and time to derive approximately equivalent algebraic equations for each node.

Solving the resulting equations by a suitable matrix inversion or iterative technique,

These steps are further explained in the next section with more details.

Example

Apply Taylor's series expansion to the form

$$\left(\frac{\partial^4 T}{\partial x^4}\right)_{i,j} = aT_{i+2,j} + bT_{i+1,j} + cT_{ij} + dT_{i-1,j} + eT_{i-2,j}$$

And hence show that

$$\left(\frac{\partial^4 T}{\partial x^4}\right)_{i,j} = \frac{T_{i+2,j} - 4T_{i+1,j} + 6T_{ij} + 4T_{i-1,j} + T_{i-2,j}}{\Delta x^4} - \frac{(\Delta x)^2}{6}\left(\frac{\partial^6 T}{\partial x^6}\right)_{i,j}$$

Solution

Writing down the Taylor' series expansions

$$T_{i\pm2,j} = T_{ij} \pm (2\Delta x)\left(\frac{\partial T}{\partial x}\right)_{i,j} + \frac{(2\Delta x)^2}{2}\left(\frac{\partial^2 T}{\partial x^2}\right)_{i,j}$$

$$\pm \frac{(2\Delta x)^3}{6}\left(\frac{\partial^3 T}{\partial x^3}\right)_{i,j} + \frac{(2\Delta x)^4}{24}\left(\frac{\partial^4 T}{\partial x^4}\right)_{i,j}$$

$$\pm \frac{(2\Delta x)^5}{120}\left(\frac{\partial^5 T}{\partial x^5}\right)_{i,j} + \frac{(2\Delta x)^6}{720}\left(\frac{\partial^6 T}{\partial x^6}\right)_{i,j} \pm \ldots\ldots$$

and

$$T_{i\pm1,j} = T_{ij} \pm (\Delta x)\left(\frac{\partial T}{\partial x}\right)_{i,j} + \frac{(\Delta x)^2}{2}\left(\frac{\partial^2 T}{\partial x^2}\right)_{i,j}$$

$$\pm\frac{(\Delta x)^3}{6}\left(\frac{\partial^3 T}{\partial x^3}\right)_{i,j}+\frac{(\Delta x)^4}{24}\left(\frac{\partial^4 T}{\partial x^4}\right)_{i,j}$$

$$\pm\frac{(\Delta x)^5}{120}\left(\frac{\partial^5 T}{\partial x^5}\right)_{i,j}+\frac{(\Delta x)^6}{720}\left(\frac{\partial^6 T}{\partial x^6}\right)_{i,j}$$

Substituting these expression into the form

$$\left(\frac{\partial^4 T}{\partial x^4}\right)_{i,j}=aT_{i+2j}+bT_{i+1j}+cT_{ij}+dT_{i-1,j}+eT_{i-2,j}$$

We obtain the following equations:

$$a+b+c+d+e=0$$

$$2a+b-d-2e=0$$

$$2a+\frac{b}{2}+\frac{d}{2}+2e=0$$

$$\frac{4}{3}a+\frac{b}{6}-\frac{d}{6}-\frac{4}{3}e=0$$

$$\frac{2}{3}a+\frac{b}{24}+\frac{d}{24}+\frac{2}{3}e=1$$

Whose solution is found to be

$a=1,\ b=-4\ ,\ c=6d=-4,\ e=1$

$$T_{i+2,j}-4T_{i+2,j}+6T_{i,j}-4T_{i-1,j}+T_{i-2,j}=(\Delta x)^4\left(\frac{\partial^4 T}{\partial x^4}\right)_{i,j}+\frac{\Delta x^6}{6}\left(\frac{\partial^6 T}{\partial x^6}\right)_{i,j}+\ldots.$$

Which immediately gives

$$\left(\frac{\partial^4 T}{\partial x^4}\right)_{i,j}=\frac{T_{i+2,j}-4T_{i+2,j}+6T_{i,j}-4T_{i-1,j}+T_{i-2,j}}{\Delta x^4}-\frac{(\Delta x)^2}{6}\left(\frac{\partial^6 T}{\partial x^6}\right)_{i,j}$$

EXPLICIT METHODS

These are basic finite difference methods for solving partial differential equations. Here, our approach in solving such equations is to replace the derivates of given differential equation by their finite difference formulae.

A formula which expresses one unknown value at a given node in terms of the known preceding values is called an explicit formula .Some of the well-known explicit methods are presented below for solving parabolic partial differential equations.

Schmidt Method

This is a simple two- level explicit method. For illustration let us consider a one-dimensional diffusion equation at the (i,n) grid point. Thus, we have

$$\left(\frac{\partial T}{\partial t}\right)_i^n = \alpha\left(\frac{\partial^2 T}{\partial x^2}\right)_i^n$$

Introducing forward difference approximation to the Time derivative and Central difference approximation to space derivative (FTCS), the above equation becomes

$$\frac{T_i^{n+1} - T_i^1}{\Delta t} = \alpha\frac{T^n{}_{i+1} - 2T_i^n + T_{i-1}^n}{(\Delta x)^2}$$

That is,

$$T_i^{n+1} = T_i^n + \frac{\alpha\Delta t}{(\Delta x)^2}\left(T_i^{n+1} - 2T_i^n + T_{i-1}^n\right)$$

We now define

$$\frac{\alpha\Delta t}{(\Delta x)^2} = r$$

Then the above equation becomes

$$T_i^{n+1} = rT_{i-1}^n + (1-2r)T_i^n + rT_{i+1}^n$$

This is the Schmidt explicit formula which gives the approximate value rT_i^{n+1} at the $(n+1)$ the time level. This is also called a two-level explicit formula.

For $n=0,$ Eq. gives approximate values of T_i^1 at grid points along the first time row $t=\Delta t,$ in terms of known initial and boundary value T_i^0 $(i=1,2,\ldots,I)$ Taking $n=1$m we can compute approximate values of T_i^2 at the grid points along the second time level, in terms of just computed values T_i^1, and so on. This process is continued until a predetermined time is reached. We shall illustrate this method through the following examples.

Example

Solve the following initial boundary value problem using an explicit finite difference method:

$$\frac{\partial T}{\partial t}=\frac{\partial^2 T}{\partial x^2} \quad 0\le x\le 1$$

Given

$$T=\sin \pi x \quad when\, t=0, 0\le x\le 1$$

$$T=0 \quad at\, x=0 \;\, and\, x=1 \;\, for\, t>0$$

And hence examine the accuracy of the numerical solution at t $=0.006$ with its analytical solution.

Solution

Let us choose $\Delta x=0.1$. To get a numerically stable solution, we choose Δt in such a way as to satisfy the stability condition

$$r=\frac{\Delta t}{\Delta x^2}\le 0.5$$

Which gives $\Delta t\le 0.005$. in this example, we take $\Delta t=0.002$, so that r = 0.2. Thus using the FTCS approximation, the given differential equation reduce to

$$T_i^{n+1}=rT_{i-1}^n+(1-2r)T_i^n+rT_{i+1}^n$$

Note that, $n=0$ corresponds t the initial condition at $i=0$ and the end values of T are given by boundary conditions. Thus,

$$T_1^0=0.3090, \qquad T_2^0=\sin\frac{\pi}{5}=0.5878, \quad T_3^0=0.8090$$

$T_4^0 = 0.9511, \quad T_5^0 = \sin\frac{\pi}{2} = 1.0, \quad T_6^0 = 0.9511$

$T_7^0 = 0.8090, \quad T_8^0 = 0.5878, \quad T_9^0 = 0.3090$

The boundary conditions give $T_0^0 = T_{10}^0 = 0.0$ Now the plan of computation, the initial and boundary conditions are shown

We observe the symmetry of the solution on either side of the centre line $x=0.5$ and therefore the value of T and $x=0.6$ is the same as x = 0.4 and so on. In other words, $T_6^0 = T_4^0\ etc.$ These entries are shown as the first row in For $i=1, n=0,$ using the numerical scheme given by Eq. we get

$T_1^1 = 0.27_0^0 + 0.6T_1^0 + 0.2T_2^0 - 0.3030$

Similarly, for $i = 2,3,4\, and\, 5, n = 0, we\, can\, T_2^1$, $T_3^1, T_4^1\ and\ T_5^1$ which is the solution of the given initial boundary value problem for $t = \Delta t = 0.002$ Thus we can use the sample algorithm described by Eq. and compute further the values of T for subsequent time steps at all the nodes, which are convenientialy tabulated in Table.

		x										
n	t	0.0	0.1	0.2	0.3	0.4	0.5	0.6	0.7	0.8	0.9	1.0
0	0.0	0.0	0.3090	0.5878	0.8090	0.9511	1.0	0.9511	0.8090	0.5878	0.3090	0.0
1	0.002	0.0	0.3030	0.5763	0.7932	0.9325	0.9804	0.9325	0.7932	0.5763	0.3030	0.0
2	0.004	0.0	0.2970	0.5650	0.7777	0.9142	0.9612	0.9142	0.7777	0.5650	0.2970	0.0
3	0.006	0.0	0.2912	0.5540	0.7625	0.8963	0.9424	0.8963	0.7625	0.5540	0.2912	0.0
		0.0	0.2912	0.5520	0.7625	0.8964	0.9425	0.8964	0.7625	0.5540	0.2912	0.0

The analytical solution to the present problem can be easily found to be see Sankara Rao,

$T(x,t) = e^{-\pi^2 t}\ sin\,\pi x$

Thus,

$T(x, 0.006) = 0.9425 \sin \pi x$

The exact solution (mark) and the computed solution at t = 0.006 is shown in Table and both solutions found to agree up to third decimal. Place. This result demonstrate the power of even simple explicit numerical method.

Example 3. Solve the differential equation

$$\frac{\partial T}{\partial t} = \frac{\partial^2 T}{\partial x^2}, \qquad 0 \le x \le \frac{1}{2}$$

Given that $T = 0$ when $t = 0, 0 \le x \le \frac{1}{2},$ and with boundary conditions

$$\frac{\partial T}{\partial x} = 0 \ at\, x = 0 \ and \frac{\partial T}{\partial x} = 1 \ at\, x = \frac{1}{2} \ for\, t > 0$$

Taking $\Delta x = 0.1 \,, \Delta t = 0.001 . Given$ solution for three time steps.

Solution. Using FTCS approximation, the discredited equation of the given partial differential equation for each node i is given by

$$T_i^{n+1} = r\,T_{i-1}^n + \left(1 - 2r\right) T_i^n + T_{i+1}^n$$

Where $r = \Delta t \,/\, \Delta x^2$. The data of present problem gives $r = 0.01.$. The require computation and the initial conditions are. In this problem, derivative boundary conditions are specified, which require special care. The implementation of which can be explained are specified as follows:

At the left end, that is at $x = 0, \partial T \,/\, \partial x = 0 \; for\, t > 0.$ Using forward difference approximation, it can be equivalently written as $\frac{T_{i+1}^n - T_i^n}{\Delta x} = 0$ implying $T_1^n = T_{i+1}^n,$ Therefore, $T_0^n = T_1^n$

At the right end, that is, at x = ½, the boundary condition is $\partial T \,/\, \partial x = 1, t > 0.$ Using backward difference approximation, its equivalent is

$$\frac{T_1^n - T_{i-1}^n,}{\Delta x} = implying\, T_i^n = T_{i-1}^n + \Delta x,$$ Therefore, $T_5^n = T_4^n + \Delta x$

when $i = 1,$

$$T_1^{n+1} = rT_0^n + (1-2r)T_1^n + rT_2^n$$

Eliminating t_0^n from Eqs. we get

$$T_1^{n+1} = (1-r)T_1^n + rT_2^n$$

Thus, we use the following equations from
$x = 0 \, to \, 0.3, thta \, is, for \, i = 01, 2 \, and \, 3$

$$\left.\begin{array}{r} T_1^{n+1} = (1-r)T_1^n + rT_2^n \\ T_0^{n+1} = T_1^{n+1} \\ T_i^{n+1} = rT_{1-l}^n + (1-2r)T_1^n + rT_{1+l}^n, \quad for \, i = 2, 3 \end{array}\right\}$$

Also, $for \, i = 4, Eq.(1) \, becomes$

$$T_4^{n+1} = rT_3^n + (1-2r)T_4^n + rT_5^n$$

Form $Eqs.$ ...we eliminate T_5^n to get

$$T_4^{n+1} = rT_3^n + (1-r)T_4^n + \frac{\Delta t}{\Delta x}, \quad for \ i = 4$$

and Eq. gives

$$T_5^{n+1} = T_4^{n+1}$$

Thus, we use Eqs. and to compute T at $x=0.4$ and $x=0.5$ that is, for $i=4$ and 5 and Eqs. for internal nodes.

Following the procedure, the solution T at each node for three time steps are computed and depicted in Table

Table:

		x					
n	t	0.0	0.1	0.2	0.3	0.4	0.5
0	0.0	0.0	0.0	0.0	0.0	0.0	0.0
1	0.001	0.0	0.0	0.0	0.0	0.01	0.11
2	0.002	0.0	0.0	0.0	0.001	0.019	0.119
3	0.003	0.0	0.0	0.0001	0.0027	0.0272	0.1272

Dulfort- Frankel Method (1953)

This is another explicit method developed for solving diffusion equation $T_t = \alpha T_{xx}$

In this method, both time and space derivative are replaced by their central difference approximations. Thus, is replaced as

$$\frac{T_i^{n+1} - T_1^{n-1}}{2\Delta t} = \alpha \frac{T_{i-1}^n - 2T_i^n + T_{i+1}^n}{\Delta x^2}$$

On rearrange, it simplifies to

$$T_i^{n+1} - T_1^{n-1} + 2r\left(T_{i-1}^n - 2T_i^n + T_{i+1}^n\right)$$

Where

$$r = \alpha \frac{\Delta t}{\Delta x^2}$$

Further, in Eq., T_i^n is replaced by its average value at $(n+1)$th and (n-1)th time levels. Thus, if we substitute

$$T_i^n = \frac{T_i^{n+1} + T_i^{n-1}}{2}$$

Into Eq., we obtain

$$T_i^{n+1} = \frac{2r}{1+2r}\left(T_{i-1}^n + T_{i+1}^n\right) + \frac{1-2r}{1+2r} T_i^{n-1}$$

Since it involves three time levels, it is called and explicit three-level finite difference scheme. To initiate computation using Durfort-Frankle Scheme, we should know the solution at two time levels, that is when $n=0\, and\, n=1$ or $t=$ and t = Δt.

To implement this scheme , from the given initial and boundary conditions, we shall compute the solution at t = Δ, using two-level explicit method and then use Eq. to obtain the numerical solution at subsequent time steps, at all nodes covering the region of computation

Implicit Methods

These are also basic finite difference methods for solving parabolic partial differential equations. A method in which the computation of many present unknowns values necessitates the solution of a set of simultaneous equations is called an implicit method. For a given problem, the choice of a particular method depends on the computational accuracy required , the affordable computational time and storage memory. Some of the well-know implicit methods are discussed in the following sections.

Classical implicit Method

This method was developed by O'Brien et al. to solve the diffusion equation Here, we replace the time derivative is replaces by its central different approximation. Thus using BTCS approximation, the diffusion equation evaluated at ($, n+1)$ grid point namely

$$\left(\frac{\partial T}{\partial t}\right)_t^{n+1} = \alpha\left(\frac{\partial^2 T}{\partial x^2}\right)_t^{n+1}$$

Gives

$$\frac{T_i^{n+1} - T_i^n}{\Delta T} = á\left[\frac{T_{i-1}^{n+1} - 2T_i^{n+1} + T_{i+1}^{n+1}}{(\Delta x)^2}\right]$$

$$-rT_{i-1`}^{n+1} + (1+2r)_i^{n+1} T_i^{n+1} - rT_{i-1`}^{n+1} = T_1^n$$

For $i = 1.2 ,.....N-1$, the number of unknowns in $x - direction$. Thus equation us implicit in the sense that there are three unknowns values of T at $(n+1)$th time level. For $n=0$ and for $i = 1,2,.....,N-1\,Eq.$ gives $(N-1)$ simultaneous linear algebraic equations for $(N-1)$ unknown value of T at internal grid points, along the first time level in terms of known initial and boundary values. These equations can be solved to get the values of T_i^1, for $i = 1,2,.....,N-1$. Similarly by taking $n = 1$, we get $(N-1)$ equationss for $(N-1)$ unknown values of T at the second time level, in terms of boundary values and the recently computed values of T at the first time level, in terms of boundary values and the recently computed values of T at the first time level and so on. Thus, for example, if the values of T are known $at\, n-th$ time level. By subsisting $i = 1, 2,.....N-1$ in Eq. we get the following set of simultaneous linear algebraic equations in tridiagonal from as given below:

$$\left.\begin{aligned} (1+2r)T_1^{n+1} - rT_2^{n+1} &= d_1^n \\ -rT_1^{n+1} + (1+2r)T_2^{n+1} - rT_3^{n+1} &= d_2^n \\ -rT_2^{n+2} + (1+2r)T_3^{n+1} - rT_4^{n+1} &= d_3^n \\ -rT_{N-2}^{n+1} + (1+2r)T_{N-1}^{n+1} &= d_{N-1}^n \end{aligned}\right\}$$

where,

$$d_1^n = T_1^n + rT_0^{n+1}, \quad d_i^n = T_i^n. \quad i = 2\ ,3\ ,\ldots\ldots, N-2$$

and

$$d_{N-1}^n = T_{N-1}^n + rT_N^{n+1}$$

The subscript 0 and N correspond to the value. This tridiagonal system can be solved exactly using Crout's LU reduction technique, as explained in section.

Crank- Nicolson Method (1947)

In this method, the diffusion equation is approximated by replacing both time and space derivative by their central difference approximations (CTCS(at a point.

$$\left[i\Delta x, \left(n + \frac{1}{2} \right) \Delta t \right]$$

Which is half way between the grid points $(i,n) and (i,n+1)$ Thus, the diffusion equations can be written as

$$\left(\frac{\partial T}{\partial t} \right)_t^{n+1(1/2)} = \alpha \left[\left(\frac{\partial^2 T}{\partial x^2} \right)_i^n + \left(\frac{\partial^2 T}{\partial x^2} \right)_i^{n+1} \right]$$

Now, using central difference approximation to both the space and time derivatives, we get

$$\frac{T_i^{n+1} - T_i^n}{\Delta T} = \frac{á}{2} \left(\frac{T_{i+1}^n - 2T_i^n + T_{i-1}^n}{\Delta x^2} + \frac{T_{i+1}^{n+1} - 2T_i^{n+1} + T_{i-1}^{n+1}}{\Delta x^2} \right)$$

or

$$-\frac{r}{2} T_{i-1}^{n+1} + (1+r) - \frac{r}{2} T_{i-1}^n + (1-r) T_i^n + \frac{r}{2} T_{i+1}^n$$

For $i = 0,1,2,\ldots\ldots,N-1,N,$ where N is the number of division, into which the interval of integration is subdivided, thus, $i = 0\, and\, i = N$ correspond to the left and right boundaries respectively.

It can be observed from Eq., that the new value of temperature T_i^{n+1} is not given directly in terms of known T_i^n at one time step but is also a function of unknown temperatures at adjacent position ,and is therefore called an implicit method.

Thus, assuming that the values of T are known at the $n-th$ time level and taking $i = 1,2\ldots,N-1, Eq$ yields a set of linear simultaneous equations in terms of unknowns T_i^{n+1} as given below in a tridiagonal form.

$$\begin{bmatrix} 1+r & -\frac{r}{2} & & & \\ -\frac{r}{2} & 1+r & -\frac{r}{2} & & \\ & -\frac{r}{2} & 1+r & -\frac{r}{2} & \\ & \vdots & & & \\ & \vdots & & & \\ & -\frac{r}{2} & 1+r & -\frac{r}{2} & \\ & & -\frac{r}{2} & 1+r & \end{bmatrix} \begin{pmatrix} T_1^{n+1} \\ T_2^{n+1} \\ T_3^{n+1} \\ \vdots \\ \vdots \\ T_{N-2}^{n+1} \\ T_{N-1}^{n+1} \end{pmatrix} = \begin{pmatrix} d_1^n \\ d_2^n \\ d_3^n \\ \vdots \\ \vdots \\ d_{N-2}^n \\ d_{N-1}^n \end{pmatrix}$$

where

$$\left.\begin{aligned} d_1^n &= \frac{r}{2}T_0^n + (1+r)T_1^n + \frac{r}{2}T_2^n + \frac{r}{2}T_0^{n+1} \\ d_1^n &= \frac{r}{2}T_{i-1}^n + (1-r)T_i^n + \frac{r}{2}T_{i+1,}^n \qquad i = 2,3,\ldots,N-2 \\ d_{N-1}^n &= \frac{r}{2}T_{N-2}^n + (1-r)T_{N-1}^n + \frac{r}{2}T_N^n + \frac{r}{2}T_N^{n+1} \end{aligned}\right\}$$

all of which are known quantities. This resulting tridiagonal system can be solved exactly using Crout's reduction technique as explained in.

In this method, not only the computational effort, but also the required computer memory space can be minimized. Also, since we use the same coefficient tridiagonal matrix to compute the solution at subsequent time steps, there will be a considerable saving in computer times as well, particularly while handling higher order systems. An example follows for illustration of this method.

Example 4

Use Crank-Nicolson method to find the numerical solution of the following parabolic partial differential equation after one time step:

$$T_t = T_{xx}, \qquad 0 < x < 1$$

Subject to the initial condition

$$T(x,0) = 1, \quad 0 < x < 1$$

and the boundary conditions

$$T(0,t) = T(1,t) = 0, \;\; t \ge 0$$

Compute the solution by taking $\Delta x = 1/4$ and $\Delta t = 1/32$.

Solution

The Crank-Nicolson implicit finite difference scheme for the given differential equation is given by Eq. That is

$$-\frac{r}{2}T_{i-1}^{n+1} + (1+r)T_i^{n+1} - \frac{r}{2}T_{i+1}^{n+1} = \frac{r}{2}T_{i-1}^{n+1} + (1-r)T_i^{n} - \frac{r}{2}T_{i+1}^{n}$$

Nothing that N = 1 / $\Delta x = 4, \Delta t$ / $\Delta x^2 = 1$ /2 and $N = 0, 4$ correspond to the left and right boundaries respectively, for $i = 1, 2, 3$ and n = 0, Eq yeilds

$$-\frac{1}{4}T_0^1 + \frac{3}{2}T_1^1 = \frac{1}{4}T_2^1 = \frac{1}{4}T_0^0 + \frac{1}{2}T_1^0 + \frac{1}{4}T_2^0$$

$$-\frac{1}{4}T_1^1 + \frac{3}{2}T_2^1 = \frac{1}{4}T_3^1 = \frac{1}{4}T_1^0 + \frac{1}{2}T_2^0 + \frac{1}{4}T_3^0$$

$$-\frac{1}{4}T_2^1 + \frac{3}{2}T_3^1 = \frac{1}{4}T_4^1 = \frac{1}{4}T_2^0 + \frac{1}{2}T_3^0 + \frac{1}{4}T_4^0$$

From the boundary conditions, we get

$$T_0^n = T_4^n = 0 \;\; \textit{for all} \;\; n = 0, 1, 2 \ldots.$$

Thus, using the boundary conditions, we get

$$0+6T_1^1-T_2^1=0+2+1$$

$$-T_1^1+6T_2^1-T_3^1=1+2+1$$

$$-T_2^1+6T_3^1-0=1+2+0$$

Which in matrix notation reduces to the tridiagonal system

$$\begin{bmatrix} 6 & -1 & \\ -1 & 6 & -1 \\ & -1 & 6 \end{bmatrix}\begin{pmatrix} T_1^1 \\ T_2^1 \\ T_3^1 \end{pmatrix}=\begin{pmatrix} 3 \\ 4 \\ 3 \end{pmatrix}$$

To solve this system, we perfectively use Crout's reduction technique, where we decompose the coefficient matrix of $Eq.$ into the form

$$A=LU=\begin{bmatrix} l_{11} & 0 & 0 \\ l_{21} & l_{22} & 0 \\ l_{21} & l_{32} & l_{33} \end{bmatrix}\begin{bmatrix} 1 & u_{12} & u_{13} \\ 0 & 1 & u_{23} \\ 0 & 0 & 1 \end{bmatrix}$$

Thus, we get

$$l_{11}=6,\ l_{21}=-1,\ l_{31}=0$$

$$l_{11}u_{12}=-1\ , l_{11}u_{13}=0\ \ implying\ u_{12}=-\frac{1}{6},\ u_{13}=0$$

$$l_{21}u_{12}+l_{22}=6\ gives\ l_{22}=6-\frac{1}{6}=\frac{35}{6}$$

$$l_{31}u_{12}+l_{32}=-1\ gives\ l_{32}=-1$$

$$l_{31}u_{13}+l_{32}u_{23}+l_{33}=6\ \ or\ \ l_{33}=\frac{204}{35}$$

Now, the system can be written as

$$\begin{bmatrix} 6 & 0 & 0 \\ -1 & \frac{35}{6} & 0 \\ 0 & -1 & \frac{204}{35} \end{bmatrix}\begin{bmatrix} 1 & -\frac{1}{6} & 0 \\ 0 & 1 & -\frac{6}{35} \\ 0 & 0 & 1 \end{bmatrix}\begin{pmatrix} T_1^1 \\ T_2^1 \\ T_3^1 \end{pmatrix}=\begin{pmatrix} 3 \\ 4 \\ 3 \end{pmatrix}$$

The solution of which is given by

$$\left(T_1^1, T_2^1, T_3^1\right) = \left(\frac{11}{17}, \frac{15}{17}, \frac{11}{17}\right)$$

Similarly, we can advance the solution one time step ar a time, by solving the same tridiagonal ∂T To solve this problem by an explicit method and to get a result of same accuracy, we should chose Δt such that

$$\Delta t < \frac{1}{32}$$

Example 5

Solve the unsteady heat equation $T_t = T_{xx}, \ 0 \le x \le 4,$ subject to the initial conditions $T(x,0) = 1000,$ and the boundary conditions

$$\frac{\partial T}{\partial x}(0,t) = 0.15(T - 70), \ \frac{\partial T}{\partial x}(4,t) = 0.25(T - 70)$$

Using Crank-Nicolson finite difference method by taking, $\alpha = 0.132, \Delta x = 1.0, \Delta t = 0.893.$ Give solution after one time step.

Solution

Form the given data, we find that

$$r = \frac{\alpha \Delta t}{\Delta x^2} = 0.25$$

and the initial conditions are shown below:

$i = 1$ $i = 2$ $i = 3$ $i = 4$ $i = 5$

$x = 0$ $x = 1$ $x = 2$ $x = 3$ $x = 4$

$T_1^0 = 1000$ $T_2^0 = 1000$ $T_3^0 = 1000$ $T_4^0 = 1000$ $T_5^0 = 1000$

Using Crank-Nicolson implicit scheme, we have front from

$$-0.25T_{i-1}^{n+1} + 2.5T_i^{n+1} - 0.25T_{i-1}^{n+1} = 0.25T_{i-1}^{n} + 1.5T_i^{n} + 0.25T_{i+1}^{n}$$

Let $n = 0, i = 1, 2, \ldots, 5, then\ Eq(1)$ gives

$$\left.\begin{array}{l} 0.25T_0^1 + 0.25T_1^1 - 0.25T_2^1 = 0.25T_0^0 + 1.5T_1^0 + 0.25T_2^0 \\ 0.25T_1^1 + 0.25T_2^1 - 0.25T_3^1 = 0.25T_1^0 + 1.5T_2^0 + 0.25T_3^0 \\ 0.25T_2^1 + 0.25T_3^1 - 0.25T_4^1 = 0.25T_2^0 + 1.5T_3^0 + 0.25T_4^0 \\ 0.25T_3^1 + 0.25T_4^1 - 0.25T_5^1 = 0.25T_3^0 + 1.5T_4^0 + 0.25T_5^0 \\ 0.25T_4^1 + 0.25T_5^1 - 0.25T_6^1 = 0.25T_4^0 + 1.5T_5^0 + 0.25T_6^0 \end{array}\right\}$$

In this system of equation, $T_o \text{ and } T_6$ are fictitious points which we shall eliminate using the boundary conditions. Using central difference approximation to the left and boundary conditions, we have

$$\frac{T_2^N - T_0^n}{\Delta x} = 0.15\left(T_1^n - 70\right)$$

Thus,

$$T_0^1 = T_2^1 - 0.15\left(T_1^1 - 70\right) \; and \; T_0^0 = T_2^0 - 0.15\left(T_1^0 - 70\right)$$

Therefore,

$$0.25\left(T_0^0 + T_0^1\right) = 0.25\left(T_2^0 + T_2^1\right) - 0.0375\left(T_1^0 + T_1^1\right) + 5.25$$

Similarly, using central difference approximation to the right-end boundary condition, we have

$$\frac{T_5^N - T_4^n}{\Delta x} = 0.25\left(T_5^n - 70\right)$$

Thus,

$$T_6^1 = T_4^1 - 0.25\left(T_5^1 - 70\right) \; and \; T_6^0 = T_4^0 - 0.25\left(T_5^0 - 70\right)$$

Therefore,

$$0.25\left(T_6^0 + T_4^1\right) = 0.25\left(T_4^0 + T_4^1\right) - 0.0625\left(T_5^0 + T_5^1\right) - 8.75$$

Hence, the last of Eq. becomes

$$-0.5T_4^1 + 2.4375T_5^1 = 0.5T_4^0 + 1.5625_5^0 - 8.75$$

Using Eqs. , it can be written in matrix notation, after using the initial conditions in the form

$$\begin{bmatrix} 2.5375 & -0.5 & 0 & 0 & 0 \\ -0.25 & 2.5 & 0.25 & 0 & \\ 0 & 0 & -0.25 & 2.5 & -0.25 \\ 0 & 0 & 0 & -0.5 & 2.4375 \end{bmatrix} \begin{pmatrix} T_1^1 \\ T_2^1 \\ T_3^1 \\ T_4^1 \\ T_5^1 \end{pmatrix} = \begin{pmatrix} 1967.75 \\ 2000 \\ 2000 \\ 2000 \\ 2053.75 \end{pmatrix}$$

Applying Crout's reduction technique, this matrix equation can be rewritten as

$$\begin{bmatrix} 2.5275 & 0 & 0 & 0 & 0 \\ -0.25 & 2.4507 & 0 & 0 & 0 \\ 0 & -0.25 & 2.475 & 0 & 0 \\ 0 & 0 & -0.25 & 2.4748 & 0 \\ 0 & 0 & 0 & 0.5 & 2.3971 \end{bmatrix} \begin{bmatrix} 1 & -0.1970 & 0 & 0 & 0 \\ 0 & 1 & -0.1020 & 0 & 0 \\ 0 & 0 & 1 & -0.1010 & 0 \\ 0 & 0 & 0 & 1 & -0.0808 \\ 0 & 0 & 0 & 0 & 1 \end{bmatrix}$$

$$x \begin{pmatrix} T_1^1 \\ T_2^1 \\ T_3^1 \\ T_4^1 \\ T_5^1 \end{pmatrix} = \begin{pmatrix} 1967.75 \\ 2000 \\ 2000 \\ 2000 \\ 2053.75 \end{pmatrix}$$

Now, the solution of Eq. can be easily found as

$$\left(T_1^1 T_2^1 T_3^1 T_4^1 T_5^1\right) = \left(971.8725 \quad 996.9774 997.8171 \; 983.2880 \; 1044.2641\right)$$

This is the heat distribution after one-time step. At subsequent time steps, the right hand side values of $Eq,$ changes while the coefficient matrix remains unchanged and hence the solution of the given problem can be obtained at subsequent time steps with little effort and so on.

Weighted Average Implicit Method

To explain this method, we shall recall the heat condition equation

$T_t = \alpha T_{xx}$

In this method, we shall replace the time derivative by its backward difference approximation and space derivative by its central difference approximation. Thus, using BTCS approximation, Eq. is first evaluated at the grid point $(i .n+1)$ to get

$T_i^{n+1} - T_i^n = r\left(T_{i-1}^{n+1} - 2T_i^{n+1} + T_{i+1}^{n+1}\right)$

where

$$r = \frac{\alpha \Delta t}{x^2}$$

Now, we introduce the weighing factor λ such that $0 \le \lambda \le 1,$ and rewrite the above expression in the form

$$T_i^{n+1} - T_i^n = r\left[\lambda\left(T_{i-1}^{n+1} - 2T_i^{n+1} + T_{i+1}^{n+1}\right) + (1-\lambda)\left(T_{i-1}^n - 2T_{i-1}^n + T_{i+1}^n\right)\right]$$

As a weighted average at $(n+1)-th$ and $n-th$ levels. On rewriting the above equations, we obtain

$$-r\lambda T_{i-1}^{n+1} + (1+2r\lambda)T_i^{n+1} - r\lambda T_{i+1}^{n+1} = r(1-\lambda)T_{i-1}^n$$

From Eq. it can be observed that when $\lambda = 1, we can$ recover O' Brienet.al, formula; when $\lambda = 1/2$, we obtain Crank-Nicolson formula and when $\lambda = 0,$ we get the explicit formula represented by Eq. Hence, this weighted average method can be considered as the mist general implicit method, which combines many methods so far discussed.

The Concept of Stability

In Section and we have introduced various finite difference methods for solving parabolic partial differential equation which models heat conduction equation. All those methods essentially reduce the solution of the partial differential equation to the solution of a set of algebraic equations. We are also aware that numerical computations carried out on any computer can be accurate only to a finite number of decimal places. At each stage of calculation, some round-off error will be introduced, no matter how small. Let e be an error introduced into the calculations at each mesh point $(i,n) and \bar{u}_i^n,$ the resulting numerical solution of the differential equation

$$u_t = \alpha u_{xx}$$

With finite difference approximation. If u_i^n is the corresponding solution of the differential equation with a finite difference method, assume no round-off error is involved in each calculation. Then, we define

$e_i^n = u_i^n - \overline{u}_i^n$

Stability considerations are very important in getting numerical solution of a differential equation, using finite difference methods. The solution of the finite difference equation is finite difference methods. The solution of the exponentially as we progress from one time step to another. That is, e_i^n should remain uniformly bounded as $n \to \infty \; while \, \Delta t, \Delta x \to 0.$ There are two standard methods for investigating the growth of error. In solving the finite difference equations: matrix method and Von-Neumann Fourier series expansion method. However, we shall discuss only matrix method in the following section to show the important aspects of theoretical study of numerical stability, Consistency and convergence of the finite difference method are other important theoretical aspects, for which the reader is advised to consult advances books on the subjects.

The Matrix method

Let us consider one- dimension heat conduction problem described by the differential equation

$u_{1=}\alpha u_{xx} \quad 0 \le x \le L$

Subject to the boundary conditions

$u = 0\, at \;\; x = 0 \;\;\; and\, x = L, t \ge 0$

And the initial condition

$u = f(x),\; 0 < x < L,\; t = 0$

Using simple explicit finite difference scheme FTCS, the given differential equation for $n = 0,1,2\; ,\ldots and\; for\, i = 0,1,2,\ldots,N$ such that $\Delta x = L$/N. The subscript 0, N correspond to x = 0, L the left and right-end boundaries , where u is known for all t. Expanding Eq. $for\, i = 1,2\ldots,(N-1),$ we have using the boundary conditions $U_0^n = u_N^n = 0$ for all t, the above system written in matrix notation takes the following form

$$\begin{pmatrix} u_1^{n+1} \\ u_2^{n+1} \\ u_3^{n+1} \\ \vdots \\ u_{N+1}^{n+1} \end{pmatrix} = \begin{bmatrix} 1-2r & r & & \\ r & 1-2r & r & \\ & r & 1-2r & r \\ & & & \\ & & r & 1-2r \end{bmatrix} \begin{pmatrix} u_1^n \\ u_2^n \\ u_3^n \\ \vdots \\ u_{N-1}^n \end{pmatrix}$$

Equivalently $U^{n+1} = [A]U^n$ where U is a column vector [A] is an (N-1) x (N-1) square matrix, which happened to be a tridiagonal matrix. Here, [A] is called an amplification matrix, U^0 is a vector of initial values, with which we have initiated our computation, The Successive rows of computation with each time step is given as

$$U^1 = [A]U^0$$

$$U^2 = [A]U^1 = [A][A]U^0 = [A^2]U^0$$

$$\vdots$$

$$U^{n+1} = [A]U^n = [A^2]\,U^{n-1} = \cdots . = [A^{n+1}]\,U^0$$

Now, suppose we have introduced errors at every nodes along n = 0 row or at t = 0 and start the computation with the vector of values $\overline{U^0}$, we then compute successive rows as follow:

$$\overline{U^1} = [A]\overline{U^0}$$

$$\overline{U^2} = [A]\,\overline{U^1} = [A^2]\,U^0$$

$$\vdots$$

$$\overline{U^{n+1}} = [\mathrm{A}]\mathrm{U}^n = [\mathrm{A}^2]\,\mathrm{U}^{n-1} = \cdots . = [\mathrm{A}^{n+1}]\mathrm{U}^0$$

Then the error vector e can be written as

$$e^{n+1} = U^{n+1} - \overline{U^{n+1}} = \left[A^{n+1}\right]\left(U^0 - \overline{U^0}\right) = \left[A^{n+1}\right]e^0$$

Indicating that the formula for error propagation is similar to the one used in the computation of U, Since the difference equation is linear, superposition principle is applicable and hence it is enough to consider the propagation of only one line of errors. A finite difference scheme is known to be stable if e^n remains bounded as $n \to \infty$, which is ensured, provided that the eigenvalue with largest modulus of the amplification matrix [A] is less than or equal to unity.

Alternatively

$$[A]=[I_{n-1}]+r[D_{N-1}]$$

Where $[I_{N-1}]$ $is\, a$ unit matrix of order (N-1) and $[D_{N-1}]$ $is\, an(N-1)$ tridiagonal matrix. We may recall linear algebra that the eigenvalue of an $N\,xN$ tridiagonal matrix are given by the formula

$$\lambda_s = B + 2\sqrt{ca}\cos\frac{s\pi}{N+1}, \quad s = 1,2.....,N$$

Thus, the eigenvalue of $[D_{N-1}]\, are$

$$\lambda_s = -2+2\cos\frac{s\pi}{N} = -4\sin^2\frac{s\pi}{2N}, \; s=1,2,\ldots,N$$

Thus, the eigenvalue are

$$1+r\left(-4\sin^2\frac{s\pi}{2N}\right)$$

Hence, the condition for the stability of simple explicit scheme to solve heat conduction equation is

$$\left|1-4r\sin^2\frac{s\pi}{2N}\leq 1\right|$$

Therefore,

$$-1\leq -4r\sin^2\frac{s\pi}{2N}\leq 1$$

Which gives the limiting value of r as

$$-1\leq 1-4r\sin^2\frac{s\pi}{2N}$$

$$r\leq \frac{1}{2\sin^2\frac{s\pi}{2N}}$$

Thus, the explicit scheme is stable if

$$\left(r=\alpha\frac{\Delta t}{\Delta x^2}\leq\frac{1}{2}\right)$$

Example 6 Show that the Crank- Nicolson scheme for the diffusion equation

$T_t = T_{xx},\ \ 0 \le x \le L$

Satisfy the conditions

$T = 0\ \ at\, x = 0,\ \ T = L\ \ for\, t > 0$

and $T = f(x),\ 0 \le x \le L\ \ for\, t = 0$ has unrestricted stability .

Solution

We may recall the Crank-Nicolson finite difference scheme, for the given diffusion equation, given by Eq

$$-rT_{i-1}^{n+1} + 2(1+r)T_{i+1}^{n+1} - rT_{i+1}^{n+1} = T_{i-1}^{n} + 2(1-r)T_i^n + rT_{i+1,}^n \quad i = 0,1\ ,2\ ,\ldots\ldots,N$$

$i = 0, N$ correspond to the left and right boundaries respectively . For $i = 1,2,\ldots.,(N-1), Eq$ can be expanded and written in the following matrix form.

$$\begin{bmatrix} 2(1+r) & -r & & \\ -r & 2(1+r) & -r & \\ & & & \\ & -r & 2(1+r) & -r \\ & & -r & 2(1+r) \end{bmatrix} \begin{pmatrix} T_1^{n+1} \\ T_2^{n+1} \\ \vdots \\ T_{N-2}^{n+1} \\ T_{N-1}^{n+1} \end{pmatrix}$$

$$= \begin{bmatrix} 2(1-r) & & & \\ r & 2(1-r) & r & \\ & & & \\ & r & 2(1-r) & r \\ & & r & 2(1-r) \end{bmatrix} \begin{pmatrix} T_1^{n} \\ T_2^{n} \\ \vdots \\ T_{N-2}^{n} \\ T_{N-1}^{n} \end{pmatrix}$$

That is,

$$(2l - rD_{N-1})\bar{T}^{n+1} = (2l + rD_{n-1})\bar{T}^{n}$$

Where,

$$D_{N-1} = \begin{bmatrix} -2\ 1 \\ 1 — 2\ 11 \\ : \\ 1 - 2\ 1 \\ 1 - 2 \end{bmatrix}$$

Thus,

$$\overline{T}^{n+1} = (2l - rD_{N-1})^{-1}(2l + rD_{N-1})\overline{T}^{n}$$

Alternatively, it can be put in the form $\overline{T}^{n+1} = [A]\overline{T}^{n}, where$

$$[A] = (2l - rD_{N-1})^{-1}(2l + r\,D_{N-1})$$

For stability, the modulus of the eigenvalue of $[A]$ should each be less than unity. However, the eigenvalue of $D_{N-1}\, are$ known to be given by

$$\lambda_s = -2 + 2\cos\frac{s\pi}{N} = -4\sin^2\frac{s\pi}{2n}$$

Hence, the eigenvalue of $[A]$ are

$$\frac{2 - 4r\sin^2\dfrac{s\pi}{2N}}{2 + 4r\sin^2\dfrac{s\pi}{2N}},\ s = 1, 2 \ldots.., N-1$$

These are clearly less than one for all positive value r. This prove that Crank-Nicolson scheme for the given problem is stable for all positive values of r, so has unrestricted stability.

Method for Two-Dimensional Equation

Consider the finite difference methods of solution for the equation

$$\frac{\partial u}{\partial t} = \alpha\left(\frac{\partial^2 u}{\partial x^2} + \frac{\partial^2 u}{\partial y^2}\right)$$

In a rectangular region bounded by $\overline{R} = Rx[0 \le t \le T]$, where R $= [0 \le x, y \le 1]$, with appropriate initial and boundary conditions. Here again, the region $\overline{R}$ in (x, y, t) space is covered by a rectangular grid with side parallel to the axes, with $\Delta x = \Delta y = h$ and $\Delta t = k$ with grid spacing in the space and time directions respectively. The grid points (x, y, t) are given by $x = i\,\Delta x, y = j\Delta y, t = n\Delta t;$ where i, j, n are all integers and $i = j = 0\,corrospond$ to the origin. The function u satisfying the difference equation at the grid points is denoted by $u_{i,j}^{n}$. with thus notation , we shall extended and two dimensional parabolic.

Implicit method

Taking $\Delta x = y\Delta = h,$ if we apply Crank=Nicolson scheme to, we obtain

$$u_{i,j}^{n+1} = u_{i,j}^{n} + \frac{r}{2}\left(u_{i-1,j}^{n+1} + u_{i+1,j}^{n+1} + u_{ij-1}^{n+1} + u_{ij+1}^{n+1} - 4u_{ij}^{n+1}\right)$$

$$+\frac{r}{2}\left(u_{i-1,j}^{n} + u_{i+1,j}^{n} + u_{i\,j-1}^{n} + u_{i\,j+1}^{n} - 4u_{i,j}^{n+1}\right)$$

Where , r $= \alpha\,\Delta t / h^2$. Further, if we define $a = -r/2,\ c = (1+2r),$ and

$$b_{i,j}^{n} = u_{i,j}^{n} + \frac{r}{2}\left(u_{i-1,j}^{n} + u_{i+1,j}^{n} + u_{i\,j-1}^{n} + u_{i\,j+1}^{n} - 4u_{i,j}^{n+1}\right)$$

We can rewrite Eq. as

$$\left(au_{i-1,j}^{n+1} + au_{i,j}^{n+1} + au_{i\,j-1}^{n+1} + au_{i\,j+1}^{n+1} = b_{i,j}^{n}\right)$$

This resulting linear algebraic system is no longer tridiagonal, in view of live unknowns $u_{ij}^{n+1}, u_{i-1j}^{n+1}, u_{i+j}^{n+1}, u_{i\,j-1}^{n+1} = u_{ij+1}^{n+1} in\,Eq.$ Such a system is in general solved by an iterative by an method.

Alternate Direction implicit Method

While solving time-dependent partial differential equations, in more than one space variable , it is advantageous to break the numerical process into one of sequentially solving simpler problems Such a process is called splitting . For example, we will discover that much is to be gained by splitting a two- dimensional problem into a sequence of two one dimensional problems. This procedure leads to the development of so called alternate direction implicit (ADI) method This method was first developed by peace man and Rachford in 1955

It is a two step process involving the solution of tridiagonal system of equations along lines parallel to x and y axes at the first and second steps respectively. When (ADI) method is applied to Eq., in the first step, we use row-wise traverse, that is, we advance from $m\Delta t\, to\left(n+1\right)\Delta t$ by approximating the space derivatives using central difference approximation at th time level in the x-direction and at $n-th$ time level in the y-direction .After row wise transverse, we proceed column-wise traverse as a second step. In this case, we advance from $\left(n+1\right)$ th time level to $\left(n+2\right)th$ time level, using central difference approximation to both the space derivatives, this time at $\left(n+1\right)th$ level in the x-direction and at ($n+2)th$ time level in the y-direction. Thus, level in the x-direction and at $\left(n+2\right)th$ time level in the $y-direction.$ Thus, level in the $x-direction$ and at $\left(n+2\right)th$ time in the $y-direction.$ Thus, the resulting expression by taking $\Delta x=\Delta y=h=r=\alpha\Delta t/h^2$ can be written down as

$$u_{i,j}^{n+1}=u_{i,j}^{n}=r\left(u_{i-1,j}^{n+1}-2u_{i,j}^{n+1}+u_{i+1,j}^{n+1}\right)+r\left(u_{i\,j-1}^{n}+u_{i,j+1}^{n}-2u_{ij}^{n}\right)$$

$$u_{i,j}^{n+1}=u_{i,j}^{n}=r\left(u_{i-1,j}^{n+1}-2u_{i,j}^{n+1}+u_{i+1,j}^{n+1}\right)+r\left(u_{ij-1}^{n+2}-2u_{i,j}^{n+2}+u_{ij+1}^{n+2}\right)$$

This split pair and was first introduced by peace man and Rachford and is known as peace man-Rachford formula. Let us apply this formula to a two-dimensional problem as follows.

Example 7

A 5rectangular plate is of 8 units. Initially, all the points on the plate are at $50^0 C$. The edge are suddenly brought to the temperatures and held at these temperatures. Use ADI method to trace the history of temperature at the internal nodes spaced two units apart, assuming that heat flows only in x and y directions

Solution

The governing equation of heat transfer is given by $u_1 = u_{xx} + u_{yy}$

Taking $\Delta x = \Delta y = h,$ the discredited from of the above equation is

$$\frac{u_{i,j}^{n+1} = u_{i,j}^{n}}{\Delta t} = \frac{u_{i-1j} - 2u_{ij} + u_{i+1j}}{h^2} + \frac{u_{ij-1} - 2u_{ij} + u_{ij+1}}{h^2}$$

Let $r = \Delta t / \Delta x^2$, then we apply ADI method, where, as a first step we use row- wise traverse and Can be rearranged in the form

$$u_{i,j}^{n+1} = u_{i,j}^{n} = r\left(u_{i-1,j}^{n+1} - 2u_{i,j}^{n+1} + u_{i+1,j}^{n+1}\right) + r\left(u_{i\,j-1}^{n} + u_{i,j+1}^{n} - 2u_{ij}^{n}\right)$$

After the row-wise traverse, we proceed column-wise traverse as a second step and the corresponding finite difference form is

$$u_{i,j}^{n+1} = u_{i,j}^{n} = r\left(u_{i-1,j}^{n+1} - 2u_{i,j}^{n+1} + u_{i+1,j}^{n+1}\right) + r\left(u_{ij-1}^{n+2} - 2u_{i,j}^{n+2} + u_{ij+1}^{n+2}\right)$$

For row-wise traverse, we consider $j = 1\,,2$ and vary $i = 1,2,3.$ Thus. Eq. gives

$$u_{1,1}^{1} - u_{1,1}^{0} = r\left(u_{0,1}^{1} - 2u_{1,1}^{1} + u_{2,1}^{1}\right) + r\left(u_{1,0}^{1} - 2u_{1,1}^{0} + u_{1,2}^{1}\right)$$

$$u_{2,1}^{1} - u_{2,1}^{0} = r\left(u_{1,1}^{1} - 2u_{2,1}^{1} + u_{3,1}^{1}\right) + r\left(u_{2,0}^{0} - 2u_{2,1}^{0} + u_{2,2}^{0}\right)$$

$$u_{3,1}^{1} - u_{3,1}^{0} = r\left(u_{1,1}^{1} - 2u_{3,1}^{1} + u_{4,1}^{1}\right) + r\left(u_{3,0}^{0} - 2u_{3,1}^{0} + u_{3,2}^{0}\right)$$

$$u_{1,2}^{1} - u_{1,2}^{0} = r\left(u_{0,2}^{1} - 2u_{1,2}^{1} + u_{2,2}^{1}\right) + r\left(u_{1,1}^{0} - 2u_{1,2}^{0} + u_{1,3}^{0}\right)$$

$$u_{2,2}^{1} - u_{2,2}^{0} = r\left(u_{1,2}^{1} - 2u_{2,2}^{1} + u_{3,2}^{1}\right) + r\left(u_{2,1}^{0} - 2u_{2,2}^{0} + u_{2,3}^{0}\right)$$

$$u_{3,2}^{1} - u_{3,2}^{0} = r\left(u_{2,2}^{1} - 2u_{3,2}^{1} + u_{4,2}^{1}\right) + r\left(u_{3,1}^{0} - 2u_{3,2}^{0} + u_{3,3}^{0}\right)$$

Which can be written in matrix notation as

$$\begin{bmatrix} 1+2r & -r & & & & \\ -r & 1+2r & -r & & & \\ -r & 1 & + & 2r & & \\ 1 & +2r & & -r & & \\ -r & 1 & + & 2r & & -r \\ & -r & 1 & +2r & & \end{bmatrix} \begin{pmatrix} u_{1,1}^1 \\ u_{2,1}^1 \\ u_{3,1}^1 \\ u_{4,1}^1 \\ u_{5,1}^1 \end{pmatrix} = \begin{pmatrix} ru_{0,1}^1 + ru_{1,0}^1 + (1-2r)u_{1,1}^1 + ru_{1,2}^1 \\ ru_{2,0}^1 + (1-2r)u_{2,1}^0 + ru_{2,2}^1 \\ ru_{0,2}^1 + ru_{1,1}^1 + (1-2r)u_{1,2}^1 + ru_{1,3}^1 \\ ru_{2,1}^1 + (1-2r)u_{2,2}^0 + ru_{2,3}^1 \\ ru_{4,2}^1 + ru_{3,1}^1 + (1-2r)u_{3,2}^0 + ru_{3,3}^0 \end{pmatrix}$$

Taking r = 1 and using the computed values in the first step, the above equations can be written in matrix notation is

$$\begin{bmatrix} 3-1 \\ -1\ 3 \\ 3-1 \\ -1\ 3 \\ 3\ -1 \\ -1\ \ 3 \end{bmatrix} \begin{pmatrix} u_{1,1}^2 \\ u_{1,2}^2 \\ u_{2,1}^2 \\ u_{3,1}^2 \\ u_{3,2}^2 \end{pmatrix} = \begin{pmatrix} -56.1904+65+53.5714+100 \\ -40.4762+25+46.4286+10 \\ -53.5714+56.1904+54.5238+90 \\ -46.4286+40.4762+48.8095+20 \\ 54.5238+53.5714+60+80 \\ -48.8095+46.4286+50+30 \end{pmatrix} = \begin{pmatrix} 162381 \\ 40.9524 \\ 147.1428 \\ 62.8571 \\ 139.0476 \\ 77.6191 \end{pmatrix}$$

Whose solution is given by

$u_{1,1}^2 = 66.0119,\ \ u_{2,2}^2 = 41.6943$

$u_{1,2}^2 = 35.6548\ \ \ u_{3,1}^2 = 61.8452$

$u_{2,1}^2 = 63.0357,\ \ u_{3,2}^2 = 46.4881$

This is the ADI method of solution at t = 8

This procedure can be continued alternatively between row-wise and column-wise traverse steady safe solution is obtained This ADI technique is more convenient for computer adaptation than for hand computation.

REFERENCES

1. *A Survey of Matrix Theory and Matrix Inequalities* by M Marcus & H Minc, Prindle, Weber & Schmidt, 1964 / Dover, 1992
2. *Matrix Analysis* and *Topics in Matrix Analysis* by R A Horn & C R Johnson, CUP 1990/1994,
3. *Matrix Computations* by G.H.Golub & C.F.Van Loan, John Hopkins University Press, 1983 ISBN 0-946536-00-7/05-8
4. *Matrix Methods in Stability Theory* by S.Barnett and C.Storey, Nelson, 1970.
5. www.wikipedia.com/Matrices

INDEX